AN OFFICIAL ETS® PUBLICATION

A GUIDE TO THE NTE

BIOLOGY AND GENERAL SCIENCE SPECIALTY AREA TEST

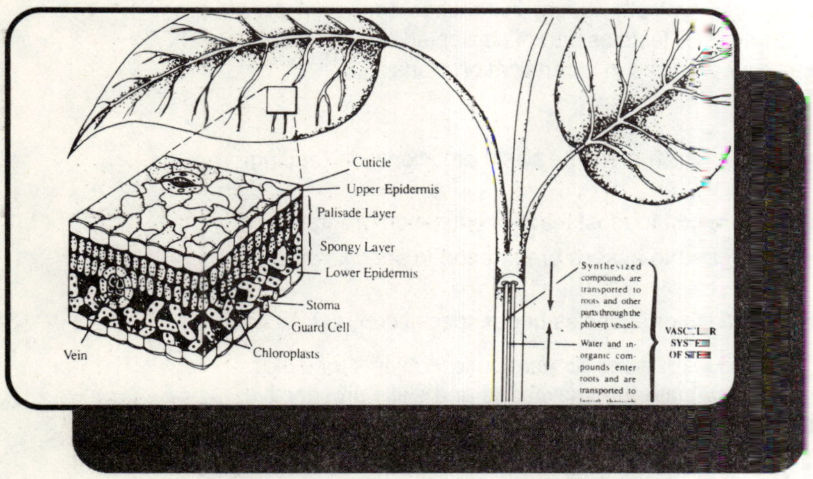

THE ONLY PRACTICE AND REVIEW BOOK WITH AN ACTUAL EXAMINATION

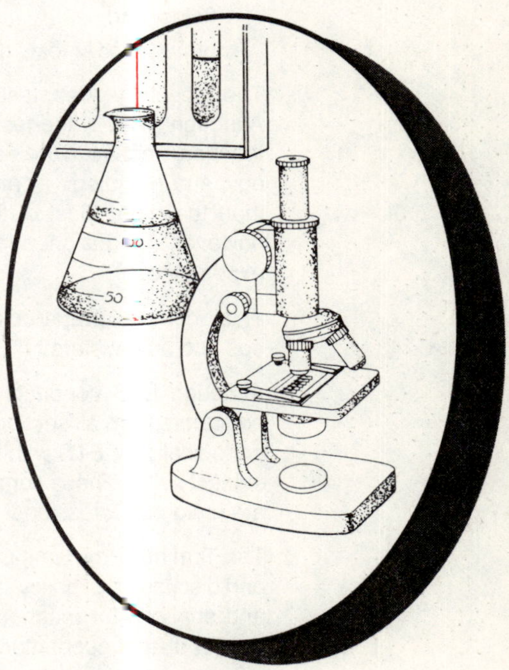

- **PREVIOUSLY ADMINISTERED NTE BIOLOGY AND GENERAL SCIENCE SPECIALTY AREA TEST**
- **INCLUDES INSTRUCTIONS FOR TEST, PRACTICE ANSWER SHEET, AND ANSWER KEY**

PREPARED BY EDUCATIONAL TESTING SERVICE

NATURE AND PURPOSE OF NTE PROGRAMS

Current NTE® tests aim to measure in an objective and standardized manner academic knowledge and skills important for beginning teachers and other education professionals. Information about academic knowledge and skills of the type gained from NTE tests is significant but limited. It is significant because knowledge and skills gained through academic preparation for a profession are obviously related to the future professional's performance. The information is limited because the academic knowledge and skills tested do not include many other elements important to professional performance, such as dedication and motivation, human relations skills, perseverance, caring, and those teaching skills learned only through practice in classroom settings. Thus, at best, NTE tests can properly be said to measure only a part, but a necessary part, of preparation for a profession.

The Pre-Professional Skills Tests (PPST®), the Core Battery, and the Specialty Area tests were developed based on academic knowledge and skills important for beginning teachers and have been used to determine that individuals have satisfied one of several criteria necessary to make them eligible for initial certification. (Review has shown that the Pre-Professional Skills Tests and the Communication Skills test cover academic knowledge and skills that are typically taught before or during the first two years of college and that the Professional Knowledge test covers material typically taught in teacher education programs.) In recent years, states have adopted NTE tests as part of career ladder and master teacher plans and as options for course equivalencies in recertification plans.

NTE Programs include:

- the Pre-Professional Skills Tests (PPST), which measure basic proficiency in reading, writing, and mathematics;
- the Core Battery, which covers areas common to most teacher education programs;
- the Specialty Area tests, which are designed to assess preparation in specific subject fields; and
- other teacher tests developed for institutions or agencies under special contract.

The Specialty Area tests measure knowledge of specific academic subjects or fields. Although most of the tests are designed to evaluate the knowledge and skills prospective teachers acquire in their academic preparation, some — such as Speech-Language Pathology, and Audiology — are designed to measure examinees' preparation for careers other than teaching. Most of the Specialty Area tests are intended for college seniors; a few, however, are designed for those who have had graduate training and/or other educational experience.

This book was prepared and produced by Educational Testing Service (ETS), which develops and administers NTE Programs tests.

Although ETS conducts the NTE Programs, it is assisted and advised by professional educators from all sections of the country. The tests themselves are developed and revised periodically by ETS with the assistance of committees of recognized authorities in specific subject fields. These committees are usually appointed from nominations made by appropriate national professional associations.

The Teacher Programs Council advises and assists ETS in ongoing examination, analysis, and discussion of policy issues concerning the NTE Programs tests and other ETS programs and services for teachers. The 20-member Council is comprised of classroom teachers, school district superintendents, college and university deans or faculty, chief state school officers, and knowledgeable members of the public.

TABLE OF CONTENTS

INTRODUCTION

The purpose of this guide is to help you prepare for the NTE Biology and General Science (BGS) Specialty Area test. The guide cannot teach you the content of the test or assure you a high score; the knowledge gained in your years of schooling is the most important factor in determining how well you will do on the test. What the book can do is acquaint you with question formats, help you pinpoint strong and weak areas and assist you in honing your test-taking skills. As a result, you can effectively prepare for an actual NTE administration.

The guide comprises four parts.

Part One provides an overview of the NTE Specialty Area tests and information about how they are developed. Various question formats are identified and explained. Part One also includes a section on test-taking strategies designed to help you reduce the anxiety often associated with taking a standardized test.

Part Two includes content specifications for the Biology and General Science test and provides sample questions with explanations of the best answers.

Part Three consists of a complete Biology and General Science practice test and a copy of the NTE Specialty Area answer sheet. The practice test is a disclosed (previously administered) form of the BGS test.

Part Four contains the answer key for the practice test and explanations of the answers. It also includes instructions for scoring the BGS test and computing your score.

PART ONE: AN OVERVIEW OF THE NTE SPECIALTY AREA TESTS

There are currently more than 70 NTE Specialty Area tests designed to assess knowledge and skills across a wide spectrum of disciplines. These tests measure the academic competence of prospective teachers in the areas in which they plan/intend to teach. However, some tests — such as Speech-Language Pathology, Audiology, School Guidance and Counseling, and School Psychologist — measure preparation for careers other than classroom teaching.

Designed for those preparing to teach secondary school Biology and General Science, the Biology and General Science (BGS) test was administered to more than 3,500 test takers in 1990-91.

Like other NTE Specialty Area tests, the Biology and General Science test is two hours long and contains 160 questions. Currently, all Specialty Area tests have a multiple-choice format with five or (in the case of tests in the language area) four answer choices (options). Some tests include audiotaped sections requiring test takers to select answers in response to taped stimuli. The BGS test does *not* contain an audiotaped section.

How the NTE Programs Specialty Area Tests Are Developed

Before a Specialty Area test is developed, educators and other professionals determine the need for such a test. Once that determination is made, test developers must ask some critical questions:

- Who will take the test and for what purposes?
- How much should the test takers be expected to know about the field?
- How should the test takers be able to use their knowledge?
- What kinds of questions should be used, and how many of each kind?
- How long should the test be?
- What should the difficulty level of the test be?

Answers to these questions come from a committee of examiners made up of subject matter experts from across the country who meet to discuss and develop preliminary specifications for the test.

Additional subject matter experts in the field review the preliminary specifications to make sure they are adequate and comprehensive. The specifications are made final only after the committee of examiners has carefully considered the information obtained from the reviewers.

The questions in an NTE Specialty Area test are prepared by ETS staff and outside experts. Every question is carefully reviewed by the ETS test development staff and editors to make sure it conforms to ETS standards. The test is then assembled by the ETS test development staff according to content and difficulty specifications established by the committee. ETS staff members, the committee members, and the subject matter experts review the assembled test to check for accuracy and appropriateness.

Before any test is administered, it must undergo a sensitivity review. This is a check to ensure that the questions truly reflect the diverse cultural nature of our society. No test will be administered if it contains questions that could be regarded as biased toward or against any subgroup of the population. The sensitivity review is conducted by ETS staff persons specially trained to identify and eliminate material that might be considered unfair or offensive to any group.

A rigorous statistical analysis is performed on each question to assure accuracy and fairness. Feedback from test takers and test center supervisors is also an important part of the ongoing review. Reports from test takers of questions thought to be ambiguously worded, for example, are thoroughly investigated. If a problem is found with a question, it is removed from the test before it is scored.

Ten Test-Taking Strategies

1. Take a watch to help you pace yourself through the test. Do not stay on any one question too long at the expense of others. Each question has the same weight in the scoring process.

2. Familiarize yourself with the exam format and the question types. Getting acquainted with the instructions and taking the time to work through the questions in Part Three of this book should help you make the best use of your time during the actual exam.

3. Take care to read each question. Make sure you know EXACTLY what is being asked, and be sure to read all the answer choices before you choose one.

4. Remember that the general instructions ask you to choose the BEST or most appropriate answer from among the choices you are given.

5. Pay close attention to the question format. The format may change in a given test. For example, some questions may stress words like NOT, LEAST, or EXCEPT. Carefully consider this type of question. It goes counter to the thinking involved in the BEST or MOST APPROPRIATE question type. With NOT, LEAST, or EXCEPT, you are looking for the *exception*, or the choice that is incompatible with the others.

EXAMPLE

Which of the following would NOT be a primary source for the investigation of a natural disaster?

(A) An article by a newspaper reporter who was present
(B) The memoirs of a witness
(C) A textbook account of the event
(D) A tape recording of the event
(E) A report by a police officer who was at the scene

Of the choices given, the only one that is NOT a primary source is C. So, the best answer choice is C.

A second type of special format is the "Roman numeral" format. Here is a sample question that illustrates this format:

EXAMPLE

Allies of the United States during the Second World War included which of the following countries?

I. Germany
II. Great Britain
III. Japan

(A) I only
(B) II only
(C) I and III only
(D) II and III only
(E) I, II, and III

In this type of format, you must choose the correct combination of options, and sometimes the options must be in a particular order or sequence. For this example, Roman numeral II, Great Britain, is the only correct answer. So, the best answer choice is B, II only.

6. Answer all the questions you are sure of first. Then, if time permits, go back to work on the more difficult questions. You may be surprised to find that in coming back and reading a question again it will be easier to understand, and you may be able to answer it.

7. Go ahead and guess. The Biology and General Science test is scored "rights only." That means that only questions answered correctly will count toward your score. Omitted questions and wrong answers will not count against you. Therefore, it will be to your advantage to answer every question, even if you have to guess.

8. Don't despair if you find that you do not know the answer to every question. You do not have to answer every question correctly to score well.

9. Take care in marking your answer sheet. Mark every answer clearly and boldly. Always make sure the question number on the answer sheet is the same as the question number in the test book. Erase all unintended marks completely. If the scoring machine scans two answers for a question, that question will be considered to have been omitted and will not count toward your score.

10. Never assume that any particular pattern of answer choices will or will not appear in the test (e.g., that choice A will appear most frequently or that the same answer choice cannot be correct three times in a row). The order and distribution of correct answer choices varies from test to test.

PART TWO: TEST CONTENT SPECIFICATIONS

The Biology and General Science examination is designed to measure examinees' preparation for teaching these subjects at the secondary school level. In particular, the questions are designed not only to test mastery of factual subject matter, but also to evaluate examinees' knowledge of the techniques and processes associated with the teaching of the subject matter. The 160 questions in the Biology and General Science test are distributed among eight categories as follows:

- **History, Philosophy, and Methodology of Science; Science, Technology, and Society.** Approximately 14-18 questions deal with these topics. Some questions may deal with environmental issues.
- **Molecular and Cellular Biology.** Approximately 22-26 questions are devoted to the molecular and cellular biology of prokaryotes and eukaryotes, including structure of biologically important macromolecules, structure and function of cells, cellular bioenergetics, photosynthesis and respiration, molecular biology of genes and gene function, and viruses.
- **Organismal Biology.** Approximately 31-35 questions deal with the biology of plants, animals, fungi and protists, including characteristics, structure and function, reproduction and development, regulation and control, behavior, and principles of Mendelian and classical genetics.
- **Evolution.** Approximately 16-20 questions are devoted to evolution, including mechanisms, population genetics, origin of life, systematics, and phylogeny.
- **Ecology.** Approximately 19-23 questions deal with ecology, including population dynamics, intraspecific competition, and life-history patterns; community characteristics, energetics, and interspecific relationships; ecosystems, aquatic and terrestrial ecosystems, interrelationships, biogeochemistry; and environmental issues.
- **Chemistry.** Approximately 14-18 questions are devoted to chemistry, including matter (structure, states, reactions), laboratory safety and laboratory experiences, and environmental issues related to chemistry.
- **Physics.** Approximately 14-18 questions deal with physics, including mechanics, heat, electricity and magnetism, wave motion, atomic and nuclear physics, and environmental issues related to physics.
- **Earth and Space Science.** Approximately 14-18 questions deal with earth and space science, including astronomy, geology, meteorology, oceanography, and environmental issues related to earth and space science.

Sample Questions

Following are several examples of the types of questions and content areas found in the Biology and General Science test. After each question is the explanation for choosing the best answer.

QUESTION 1

Which of the following acts as a processing, packaging, and distribution center for the substances that the cell manufactures?

(A) Mitochondrion (B) Lysosome
(C) Nucleolus (D) Golgi complex
(E) Ribosome

Explanation

A mitochondrion is an organelle in which most of the cell's energy is produced. A lysosome contains hydrolytic enzymes and is involved in breaking down macromolecules. The nucleolus is the site at which ribosomal subunits are made, and a ribosome is the site of protein synthesis. The Golgi complex processes, packs, and distributes proteins and protein-containing substances to be exported from the cell; thus, choice D is the correct answer.

QUESTION 2

Which of the following is responsible for producing the greatest amount of ATP in aerobic eukaryotic organisms?

(A) Glycolysis (B) Chemiosmosis
(C) Fermentation (D) Citric acid (Krebs) cycle
(E) Calvin cycle

Explanation

Glycolysis is the breakdown of glucose to pyruvic acid; this breakdown produces 2 molecules of ATP per molecule of glucose. Fermentation is the breakdown of organic compounds in the absence of oxygen and yields 2 molecules of ATP per molecule of glucose. The citric acid cycle is a stage in cellular respiration in which acetyl groups are broken down to CO_2 with the production of 2 molecules of ATP per molecule of glucose. The Calvin cycle is part of the process of photosynthesis and uses ATP for the production of sugar. It is during chemiosmosis that the majority of ATP is made; 34 molecules of ATP are produced per molecule of glucose. Thus, choice B is the correct answer.

QUESTION 3

$$N_2 + 3\,H_2 \longrightarrow 2\,NH_3$$

The balanced chemical equation above represents a process for the production of ammonia. What is the maximum number of NH_3 molecules that can be produced from a mixture of n molecules of N_2?

(A) n • 3/2 (B) n/2 (C) 2n (D) 4n (E) 6n

Explanation

If *n* is added as a coefficient in the equation

$$N_2 + 3\,H_2 \longrightarrow 2\,NH_3,$$

the following balanced equation is obtained.

$$n\,N_2 + 3n\,H_2 \longrightarrow 2n\,NH_3$$

Thus, choice C is the correct answer.

QUESTION 4

In a population that is in Hardy-Weinberg equilibrium, the percentage of the population homozygous recessive for attached earlobes is 4 percent. What percentage of the population will be heterozygous for this characteristic?

(A) 16% (B) 32% (C) 64% (D) 80% (E) 96%

Explanation

The Hardy-Weinberg equation can be expressed as $p^2 + 2pq + q^2 = 1$, where *p* is the frequency of free earlobes and *q* is the frequency of attached earlobes. If q^2 is 0.04, then *q* is 0.2. From $p + q = 1$, *p* is 0.8. Thus p^2 is 0.64 and $2pq$ is 0.32. The percentage of individuals heterozygous for free earlobes, therefore, is 32 percent, and choice B is the correct answer.

QUESTION 5

Within a food web, energy is transferred from one trophic layer to another. The LEAST amount of energy is available to

(A) producers (B) herbivores (C) carnivores (D) omnivores (E) decomposers

Explanation

Producers are algae or plants, and herbivores consume plants and other photosynthetic organisms for food and energy. Carnivores eat animals, and omnivores eat both plants and animals. Decomposers obtain their energy from substances such as cellulose and nitrogenous waste products. In general, only about 10 percent of the energy stored in plants is converted to animal biomass in the herbivores that eat the plants. A similar transfer of energy (10 percent) is found at each trophic level. The carnivores, choice C, would have the least amount of energy available to them.

QUESTION 6

A student is about to go on stage in his first major dramatic role. Before his entrance, he starts breathing faster, his heartbeat increases, and he becomes rather pale. Which of the following hormones is most likely causing these reactions?

(A) Testosterone (B) Adrenaline (C) Insulin (D) Calcitonin (E) Oxytocin

Explanation

Testosterone stimulates development of the reproductive organs and male sexual characteristics. Insulin lowers blood sugar and increases glycogen storage in the liver. Calcitonin lowers blood calcium levels and oxytocin stimulates contraction of the uterine muscles and mammary gland cells. Adrenaline is a hormone produced by the adrenal gland in response to fear and apprehension. It increases the concentration of blood sugar, raises blood pressure and the rate of heartbeat, dilates and constricts specific blood vessels, and increases muscular power and resistence to fatigue. The student is showing the typical effects of adrenaline in that he is pale and his heartbeat and breathing are faster. Thus, choice B is the correct answer.

QUESTION 7

The chlorofluorocarbons are major contributors to which of the following?

(A) Disruptions in the carbon cycle

(B) Decreases in nitrogen fixation in chemically fertilized fields

(C) The depletion of the ozone layer

(D) An increase in the pH of rainwater

(E) An increase in the macronutrient concentration of lakes

Explanation

The chlorofluorocarbons are used in refrigeration systems and fire extinguishers and were for a long time used as propellants in aerosol cans. When released into the atmosphere, these chemicals rise to the ozone layer. There, in the presence of ultraviolet light, they react with the ozone and break it down. Thus, choice C is the correct answer.

QUESTION 8

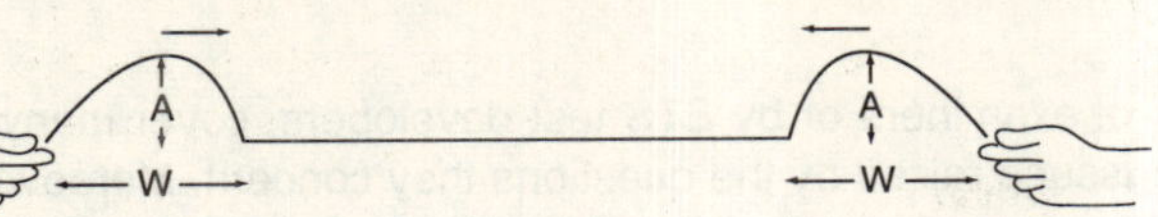

Two students holding the opposite ends of a rope flick their respective ends to produce identical wave pulses traveling toward each other along the rope. If the two pulses each have amplitude A and width W, which of the following diagrams correctly shows the resultant waveform at the center of the rope at the instant when the two pulses completely superpose?

(A)

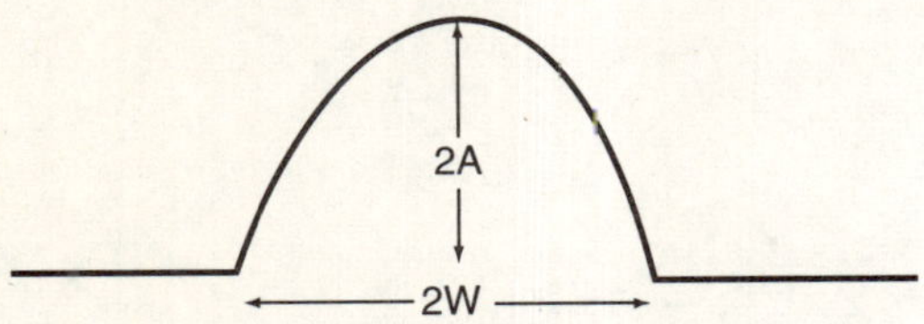

(B)

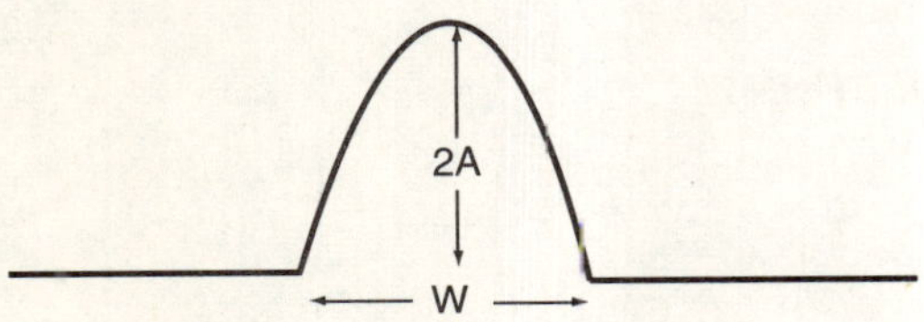

(C)

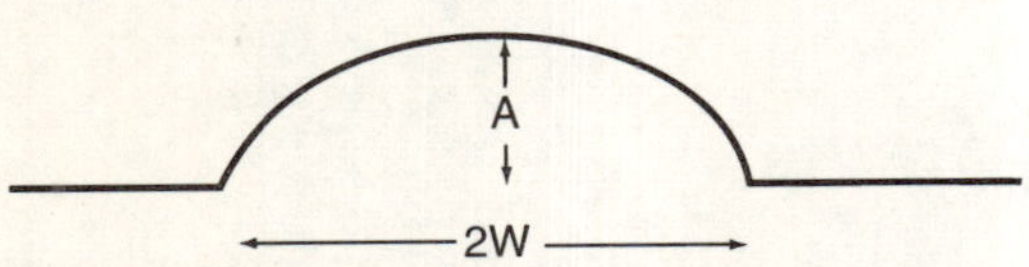

(D)

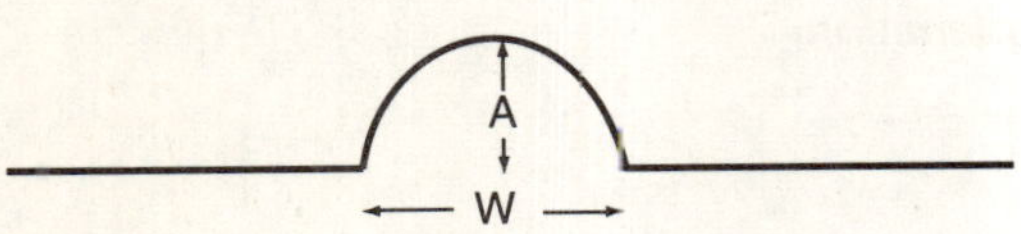

(E)

Explanation

Each of the pulses represents a transverse wave traveling as a crest along the rope. At the instant that the two pulses completely superpose, their respective transverse displacement will combine to form a resultant waveform of amplitude 2A. The width of the resultant wave, however, will remain the same as the width of the individual pulses. Thus, choice B is the correct answer. Choice E would be possible only if the two identical pulses were such that one was a crest and the other was a trough.

QUESTION 9

Surface temperatures in the temperate zones of the Northern Hemisphere are higher in the summer months (June to August) than in the winter months (December to February) because the

(A) distance from the Earth to the Sun is greatest in December

(B) distance from the Earth to the Sun is least in December

(C) angle of the Sun's rays with the Earth's surface is closest to 90° in June

(D) angle of the Sun's rays with the Earth's surface is farthest from 90° in June

(E) atmosphere is more often clouded in the winter than it is in the summer

Explanation

Temperatures on Earth change in an annual cycle because the Earth is tilted at an angle of 23.5 degrees on its axis as it travels around the Sun. In the Northern Hemisphere in summer, the Earth is tilted toward the Sun, and thus the angle at which the Sun's rays strike the surface of the Earth is closest to 90 degrees. In the winter, the North Pole is tilted away from the Sun and the angle of the Sun's rays when they strike the surface of the Earth is farthest from 90 degrees. The angle of the Sun's rays with the Earth's surface, therefore, is closest to 90 degrees in June, and choice C is the correct answer.

QUESTION 10

The complementarity of one strand of DNA with the other strand was explained by

(A) Hershey and Chase

(B) Watson and Crick

(C) Griffith

(D) Miescher

(E) Chargaff

Explanation

All of the scientists listed in the options contributed to the discovery of DNA and to the elucidation of its structure. Miescher isolated DNA from the nuclei of cells and Griffith discovered transformation. Hershey and Chase demonstrated that DNA is the hereditary material of a virus. Chargaff found that the proportion of the four nitrogen bases is the same in all cells of all individuals of a given species and that there is always approximately the same amount of adenine as thymine and guanine as cytosine in a nucleus. Watson and Crick discovered the structure of DNA and explained the complementarity of one strand of DNA with the other. Thus, choice B is the correct answer.

PART THREE: PRACTICE TEST

Now that you have been introduced to the content of the test, have considered some test-taking strategies, and have read through sample questions, you should try taking the practice test. You will probably find it helpful to simulate actual test conditions, following the directions exactly and giving yourself two hours to work on the questions. An answer sheet can be found at the end of Part Three. When you have finished the test, you can turn to the instructions for scoring your test in Part Four.

You will find it useful to read the explanations of the best answer choices for the questions in the practice test. These explanations, which were written by members of the committee of examiners or by ETS test developers, cover many of the issues raised by the questions they concern. Moreover, they present a fair sample of the kinds of questions asked and issues raised in the Biology and General Science Specialty Area test.

Keep in mind that the test you take at an actual administration may be slightly more difficult or slightly easier than the test included in this book. The content of test questions may be different, and you yourself may not perform at exactly the same level every time you take a test. Therefore, you should not expect to get the same score when you take a test at an actual administration that you get when you take the practice test.

Print your full name here: ______________________________
(last) (first) (middle)

FORM 3NNT1

03

Specialty Area Test

NTE® Programs

Biology and General Science

ETS®

TEST CODE

03051

Read the directions on the back cover.

Do not break the seal until you are told to do so.

*Hand in this test book and the answer sheet **separately** at the conclusion of the test.*

THIS TEST BOOK MUST NOT BE TAKEN FROM THE ROOM.

BIOLOGY AND GENERAL SCIENCE

Time—120 minutes

160 Questions

Directions: Each of the questions or incomplete statements below is followed by five suggested answers or completions. Select the one that is best in each case and then fill in the corresponding lettered space on the answer sheet with a heavy, dark mark so that you cannot see the letter.

1. The release of an egg from the ovary is called

(A) menstruation
(B) ovulation
(C) oogenesis
(D) cleavage
(E) fertilization

2. Sound waves travel fastest through a

(A) vacuum
(B) gas at high pressures
(C) gas at low pressures
(D) solid
(E) liquid

3. In evolutionary history, which of the following classes of organisms is believed to have preceded the others?

(A) Osteichthyes
(B) Hydrozoa
(C) Aves
(D) Reptilia
(E) Insecta

4. When white light hits a black object, the light is primarily

(A) reflected
(B) scattered
(C) absorbed
(D) refracted
(E) diffracted

5. The phases of the Moon are mainly due to the

(A) distance of the Moon from the Earth
(B) distance of the Moon from the Sun
(C) rotation of the Earth on its axis
(D) angle of the Moon's orbit to that of the Earth's orbit
(E) position of the Earth and Moon relative to the Sun

6. The attractive force that most strongly influences or governs the movement of astronomical bodies is called the

(A) cohesive force
(B) magnetic force
(C) gravitational force
(D) centripetal force
(E) electrical force

7. Which of the following is an advantage of using wood, rather than fossil fuels, to heat or supplement the heating of homes?

(A) Wood is a renewable natural resource.
(B) Wood provides more heating value, kilogram for kilogram, than coal, oil, or natural gas does.
(C) Wood is easier for people to obtain, transport, and store than fossil fuels are.
(D) Only wood contains compounds of hydrogen and carbon, which provide heating value.
(E) Wood does not produce air pollution.

8.

Time (hours)	Observed Counts per Minute
0	204
1	106
2	51
3	22
4	13

A radioactive source is monitored for one minute at hourly intervals by use of a Geiger counter, yielding the data above. All of the following statements are consistent with these data EXCEPT:

(A) Statistical fluctuations are present in the data.
(B) Doubling the observation times to two minutes will approximately double the number of counts recorded.
(C) The half-life of the source is approximately one hour.
(D) The count rate will increase in the coming hours.
(E) An hour before the first observation, approximately 400 counts would have been recorded during a one-minute observation.

GO ON TO THE NEXT PAGE.

9. Approximately how many years will it take for light from Sirius, the Dog Star, to reach Earth if Sirius is about 8×10^{13} kilometers distant from Earth? (1 light-year = 9.46×10^{12} kilometers)

(A) 0.85 years
(B) 1.2 years
(C) 8.5 years
(D) 12 years
(E) 75 years

10. Prior to the appearance of photosynthetic organisms, the early atmosphere of the Earth is believed to have contained only trace amounts of

(A) N_2
(B) O_2
(C) NH_3
(D) CH_4
(E) H_2O

11. The addition of fertilizers to a lake will initially affect the oxygen cycle in the lake by

(A) promoting algal growth
(B) breaking down oxygen directly
(C) causing more oxygen from the air to be used at the lake's surface
(D) causing organisms to use less oxygen
(E) inhibiting oxygen production, causing an increase in carbon dioxide production

12. All of the following represent examples of chemical changes EXCEPT

(A) maple leaves photosynthesizing
(B) wax melting
(C) wood burning
(D) silver tarnishing
(E) iron rusting

13. Which of the following would be most useful for separating blood cells from plasma in order to study each component separately?

(A) Electron microscope
(B) Stereomicroscope
(C) Microdissection instruments
(D) Filter paper
(E) Centrifuge

14. Translation directly involves all of the following EXCEPT

(A) rRNA
(B) tRNA
(C) mRNA
(D) amino acids
(E) DNA

15. The Galapagos finches studied by Darwin are a good example of

(A) inheritance of acquired characteristics
(B) nonselection
(C) adaptive radiation
(D) speciation by polyploidy
(E) convergent evolution

16. When a homozygous red-flowering snapdragon is crossed with a homozygous white-flowering snapdragon, all the offspring produce pink flowers. This phenomenon is known as

(A) incomplete dominance
(B) sex-linkage
(C) pleiotropy
(D) recombination
(E) inversion

17. In a rocky region with no soil, the first stage of organisms in an ecological succession is usually

(A) mosses and ferns
(B) oaks and pines
(C) lichens and mosses
(D) oaks and beeches
(E) ferns and shrubs

18. A recently captured young elephant trumpets loudly and sways from side to side when a person enters its stockade. This behavior continues but diminishes in intensity until it finally disappears. This change in behavior can be classified as

(A) habituation
(B) trial and error (operant conditioning)
(C) imprinting
(D) classical conditioning
(E) imitative learning

GO ON TO THE NEXT PAGE.

19. The superficial resemblance between a porpoise and a fish is due to

(A) adaptive radiation
(B) convergent evolution
(C) nonrandom mating
(D) gene migration
(E) genetic drift

20. All of the following are primary functions of the mammalian skeleton EXCEPT

(A) movement
(B) storage of fat and inorganic salts
(C) support and shape to the body
(D) heat regulation
(E) protection

21. As a stream flows downhill, it carries a load of sediments. If the speed of the water decreases, which of the following occurs?

(A) The amount of erosion by the stream increases.
(B) The kinetic energy of the stream increases.
(C) The potential energy of the stream increases.
(D) Deposition of sediment by the stream increases.
(E) The particle size of sediment carried by the stream increases.

22. During a classroom discussion on the evolution of the giraffe's neck, a student remarked, "The giraffe stretched its neck while reaching for higher leaves, and its offspring had longer necks as a result." To correct the student's misconception, the teacher should emphasize which of the following?

(A) Overproduction of offspring leads to a struggle for survival.
(B) Only favorable adaptations have survival value.
(C) Spontaneous mutations can cause the sudden appearance of new traits.
(D) Disuse of an organ will often lead to its disappearance from a species.
(E) Characteristics acquired during one's lifetime are not passed on through the genes.

23. Houses have been built in a desert area. Trees, lawns, and flowers have been established with the aid of artificially supplied water. Which of the following would most likely occur if water were no longer supplied to the area?

(A) The area will go through secondary succession and the result likely will be a chaparral.
(B) The area will go through secondary succession and the result likely will be a deciduous forest.
(C) The land will return to desert.
(D) A grassland will form since the lawn grasses will remain after the trees die from a lack of water.
(E) Nothing will happen because houses are present.

24. A substance that contains atoms with unstable nuclei is said to be

(A) ionic
(B) covalent
(C) radioactive
(D) amphoteric
(E) electronegative

25. Scientists have proposed in recent years that the burning of fossil fuels increases the amount of CO_2 in the atmosphere. It has been predicted that if the level of CO_2 continues to rise, then the

(A) Earth's surface will have increased exposure to ultraviolet rays
(B) average annual surface temperature will increase enough to cause major ecological disruptions
(C) excess CO_2 concentration when inhaled will cause an adverse effect on animals
(D) Earth will no longer be able to absorb enough solar radiation to warm its surface
(E) ozone layer will not be able to absorb all of the extra CO_2

26. In a large population of organisms that is in Hardy-Weinberg equilibrium, gene A is found in two allelic forms, A and a. If the frequency of the dominant allele A in the population is 0.7, approximately how many individuals out of a hundred would be expected to be homozygous recessive?

(A) 6
(B) 9
(C) 14
(D) 30
(E) 49

GO ON TO THE NEXT PAGE.

27. Under the same environmental conditions, a rock broken into small pieces weathers faster than the same rock remaining in one large piece. This phenomenon is best accounted for by which of the following?

(A) The hardness of the rock is greater than the hardness of the pieces.
(B) The pieces have a greater total surface area than the large rock has.
(C) The heat generated from breaking the rock causes a chemical change in the pieces.
(D) Rocks in water weather faster than rocks on land do.
(E) All of the pieces together weigh less than the large rock does.

28. DNA molecules of different sizes can be separated in the laboratory by use of

(A) spectrophotometry
(B) immunofluorescence
(C) electrophoresis
(D) electron microscopy
(E) X-ray crystallography

29. Three organisms are described below.

Organism 1 is bilaterally symmetrical, coelomate, shows protostome-type development, possesses a chitinous exoskeleton, and is parthenogenetic.
Organism 2 is bilaterally symmetrical in the larval form, coelomate, shows deuterostome-type development, possesses an internal skeleton made of calcium-containing plates and a water-vascular system, and is dioecious.
Organism 3 is bilaterally symmetrical, coelomate, shows protostome-type development, possesses a hydrostatic skeleton, and is monoecious (hermaphroditic).

Which of the following statements concerning phylogenetic relationships among the organisms is most likely correct?

(A) Organisms 1 and 2 are more closely related to each other than either is to organism 3.
(B) Organisms 1 and 3 are more closely related to each other than either is to organism 2.
(C) Organisms 2 and 3 are more closely related to each other than either is to organism 1.
(D) Organisms 1, 2, and 3 are all equally closely related.
(E) There is not enough information to distinguish phylogenetically between organisms 1, 2, and 3.

30.

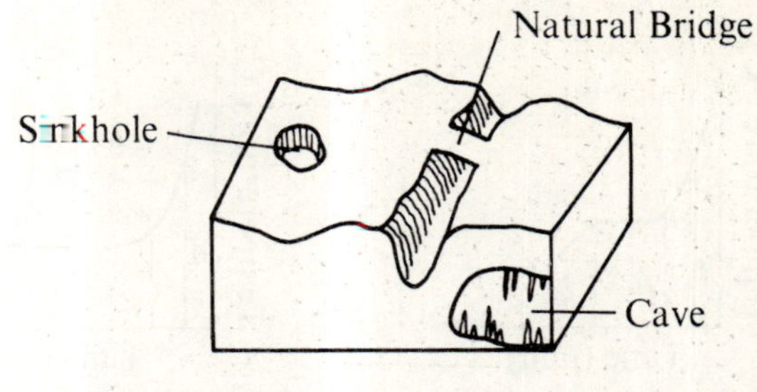

Which of the following correctly identifies the one factor that would result in the formation of the three erosional features indicated in the diagram above?

(A) Groundwater
(B) Volcano
(C) Wind
(D) Runoff
(E) Earthquake

31.

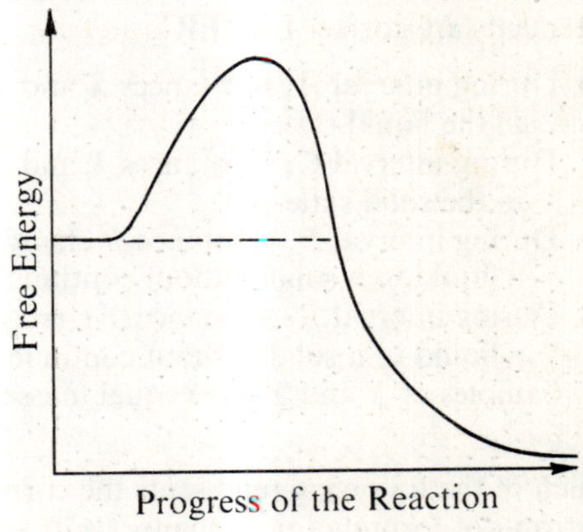

From the graph above, all of the following information concerning an enzyme reaction is available EXCEPT the

(A) energy absorbed by the enzyme
(B) energy content of the enzyme-substrate complex
(C) potential energy of the substrate molecule
(D) activation energy for the reaction
(E) energy released during the reaction

32. In the United States, most of the region east of the Mississippi River is occupied by which of the following biomes?

(A) Tundra
(B) Taiga
(C) Temperate deciduous forest
(D) Tropical rain forest
(E) Grassland

GO ON TO THE NEXT PAGE.

33.

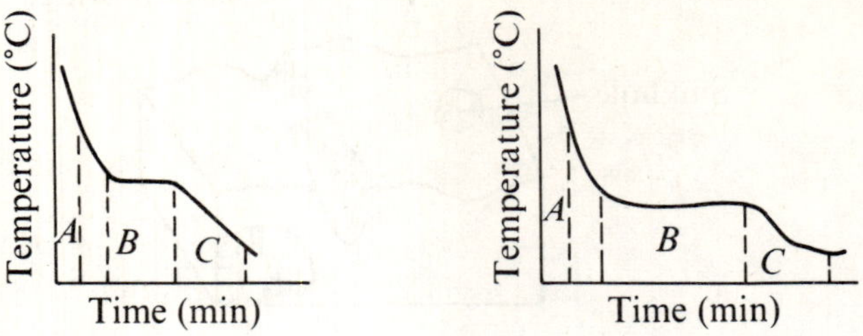

Substance *X* Substance *Y*

When the data obtained from either the cooling or melting of a substance are graphed, the presence of a plateau indicates an interval during which the liquid form of the substance is freezing or the solid form is melting without a change in temperature. The graphs above represent data from an experiment comparing the cooling of liquids *X* and *Y*. Based on the information above, all of the following statements are correct EXCEPT:

(A) During interval *A*, substances *X* and *Y* exist in the liquid state.
(B) During interval *C*, substances *X* and *Y* exist in the solid state.
(C) During interval *B*, substance *X* changes from a liquid to a solid without continued cooling.
(D) During interval *B*, substance *Y* changes from a liquid to a solid without continued cooling.
(E) Samples of *X* and *Y* have equal masses.

34. Which of the following represents the correct electron-dot formula for methane, CH_4?

(A) H:C:H:H:H

(B) H:H:C:H:H (with a pair of dots above and below C)

(C) :H:C:H:H:H: (with pairs of dots above and below each H and C)

(D)
:H:
:H:C:H:
:H:
(with pairs of dots above and below each symbol)

(E)
H
H:C:H
H
(with a pair of dots between C and each H above and below it)

35. Which of the following major scientific pronouncements was proposed first?

(A) Koch's postulates concerning diseases and their causative agents
(B) Oparin's hypothesis about the origin of life
(C) Darwin's theory of evolution
(D) Watson and Crick's model of DNA
(E) Beadle and Tatum's one gene-one enzyme hypothesis

36.

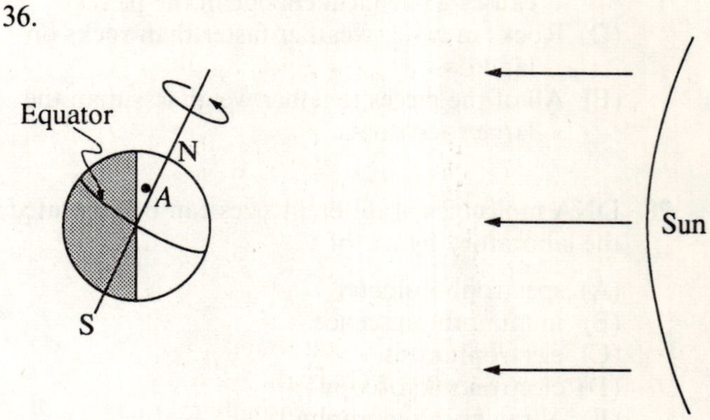

The diagram above shows one position of the Earth in relation to the Sun. When the Earth is in this position, what is the season of the year and the time of day at point *A* on the Earth's surface?

(A) Summer and morning
(B) Summer and noontime
(C) Summer and afternoon
(D) Winter and morning
(E) Winter and afternoon

37. Two organisms mate in their natural habitat and produce fertile offspring. Which of the following is the LEAST inclusive group to which these organisms belong?

(A) Genus
(B) Order
(C) Class
(D) Species
(E) Phylum

GO ON TO THE NEXT PAGE.

38. The reason that a hypothesis is easier to disprove than to prove is that

(A) formulating a hypothesis is the first step in the application of scientific thinking to a problem; disproving as a second step is easier than the first step
(B) a hypothesis can become a law only if it is proved to be true in at least a dozen different research centers; it can be disproved at a single center
(C) measurements with precision can disprove a hypothesis; measurements must have both accuracy and precision to prove a hypothesis
(D) a hypothesis can be disproved by one person but can only be developed by many scientists working together
(E) one experiment may disprove a hypothesis but many measurements cannot show that a hypothesis is always true

39. At about three years of age, male lions leave the pride in which they were born. After a period of nomadic existence, each eventually takes over a pride of its own. What is the evolutionary advantage of this behavior pattern?

(A) It increases competition between males of the same pride.
(B) It permits females to recover from their reproductive event.
(C) It enhances female receptivity to male mating behavior.
(D) It increases the likelihood of outbreeding.
(E) It encourages pair-bonding once males and females reunite.

40. If a ball is dropped from a tall tower, and at the same time a bullet is fired horizontally from the same location on the tower, the ball and bullet hit the ground at the same time.

For the statement above to be true, conditions that must be met include which of the following?

I. The ground must be level.
II. Air resistance must be absent.
III. The ball and the bullet must have the same mass.

(A) I only
(B) III only
(C) I and II only
(D) II and III only
(E) I, II, and III

41. Which of the following factors help(s) to account for the fact that, in the United States, lands to the east of a mountain range receive less rain than lands to the west of the mountains receive?

I. Winds often blow from west to east.
II. An air mass cools as it goes over a mountain range.
III. A sample of cold air can contain less water vapor than a sample of warm air can.

(A) I only
(B) III only
(C) I and II only
(D) II and III only
(E) I, II, and III

42. The gene pool of a population of *Rana catesbiana* living in a pond has remained constant and stable for many generations. Which of the following is the most likely reason for this stability?

(A) The pond ecosystem experienced rapid succession and then reached equilibrium.
(B) No mutations occurred in a large migrating population.
(C) No migration occurred in a large population that was mating randomly.
(D) Random mating occurred in a small population that had a high frequency of gene mutations.
(E) Nonrandom mating occurred in a small population that had a high frequency of gene mutations.

43. Toxic materials are often found in high concentrations in the tissues of organisms at the top of food pyramids because these materials are first

(A) degraded in plants
(B) metabolized by plants
(C) excreted as wastes by animals
(D) accumulated in the cells of autotrophs and herbivores
(E) absorbed from the abiotic environment by carnivores

GO ON TO THE NEXT PAGE.

44.

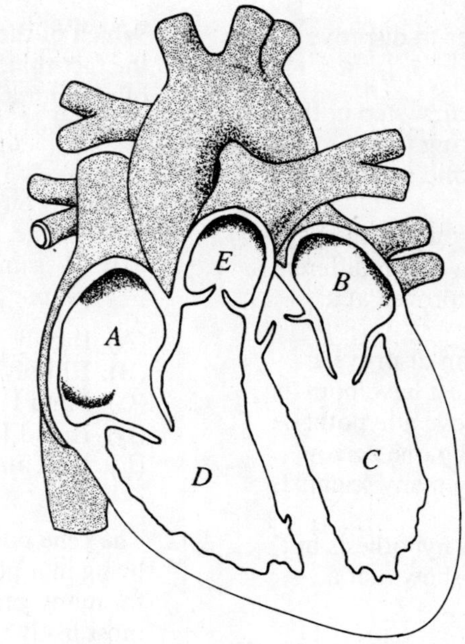

In the figure shown above, blood leaving the human heart and passing to the coronary circulatory system is expected to empty from which region?

(A) *A*
(B) *B*
(C) *C*
(D) *D*
(E) *E*

---

45. In the life cycle of the dog tapeworm, the flea, which can be assumed to be the only obligate intermediate host, gets the eggs from a dog's feces. When a dog eats the flea, the tapeworm larvae mature and mate in the intestine of the dog. Which of the following statements is correct about the eradication of the dog tapeworm?

(A) It could be accomplished by the elimination of fleas.
(B) It would require both the elimination of fleas and the ingestion of antitapeworm medication by all dogs.
(C) It would require both the ingestion of antitapeworm medication by all dogs and the sterilization of dog feces.
(D) It would require the elimination of fleas, the ingestion of antitapeworm medication by all dogs, and the sterilization of dog feces.
(E) It is impossible to achieve because of the complicated life cycle of the dog tapeworm.

46. Monarch butterflies are distasteful to blue jays. Viceroy butterflies look very much like monarchs and consequently are not eaten by blue jays. This example best illustrates the phenomenon of

(A) mimicry
(B) altruism
(C) convergent evolution
(D) genetic drift
(E) the Hardy-Weinberg principle

GO ON TO THE NEXT PAGE.

47. Which of the following methods of insect pest control is potentially the most damaging to the environment?

(A) Baiting traps with the insect pest's pheromones
(B) Applying organic broad-spectrum pesticides to the areas being protected
(C) Genetically engineering crops that are resistant or immune to the insect
(D) Using a species-specific virus to attack the pest
(E) Using radiation to sterilize laboratory-raised male insects and releasing them to compete with wild males

48. Two structures that are found in the amniotic egg and that persist in the development of placental mammals are the

(A) chorion and the placenta
(B) yolk sac and the shell
(C) allantois and the umbilical cord
(D) chorion and the amnion
(E) vitelline membrane and the amnion

49. In living organisms, the breakdown of polymers into monomers most often occurs by

(A) hydrolysis
(B) condensation
(C) dehydration
(D) anabolism
(E) photolysis

50. Which of the following best describes the mass and the weight of a woman on the Moon, compared to her mass and weight on the Earth?

	Mass on Moon Compared to Earth	Weight on Moon Compared to Earth
(A)	1/36	1/36
(B)	1/6	1/36
(C)	Less	Same
(D)	Same	More
(E)	Same	Less

51. Suppose that the arrow end of a magnetized needle is a south pole. In which of the following diagrams does the arrow indicate the correct magnetic field direction? (N and S stand for north and south magnetic poles, respectively.)

(A) 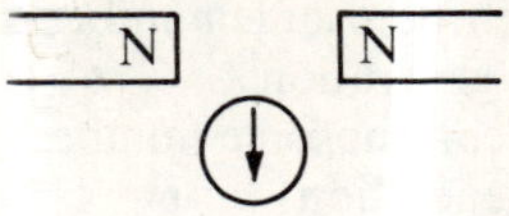

(B) 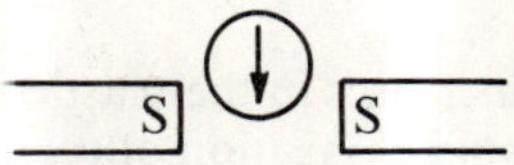

(C) 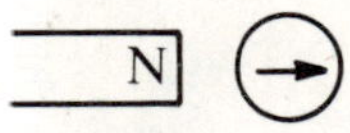

(D) 

(E) 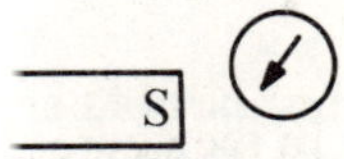

52. Although adenosine triphosphate (ATP) is continually utilized during exercise, the temperature in the body does not change appreciably. This is an example of

(A) oxidation
(B) photophosphorylation
(C) homeostasis
(D) growth
(E) glycolysis

GO ON TO THE NEXT PAGE.

53. Characteristics of both mosses and liverworts include which of the following?

(A) A dominant sporophyte generation
(B) True roots, stems, and leaves
(C) Water-dependent fertilization
(D) A diploid (2n) chromosome number in the gametophyte generation
(E) A haploid (1n) chromosome number in the sporophyte generation

54. Apical dominance, increases in the plasticity of cell walls, and the rapid elongation of cells are controlled by which of the following plant hormones?

(A) Auxins
(B) Gibberellins
(C) Cytokinins
(D) Ethylene
(E) Abscisic acid

55. The essential amino acids must be obtained from food in the human diet because these amino acids

(A) form peptide bonds more easily than other amino acids do
(B) are found in special sequences of human proteins
(C) are more common in human proteins than in other proteins
(D) are not synthesized by the human body
(E) cannot be digested by humans

56. Since color blindness is a sex-linked (X-linked) trait in humans, a man who is color-blind will pass on the allele for color blindness to

(A) half of his sons only
(B) half of his daughters only
(C) half of his sons and half of his daughters
(D) all of his sons
(E) all of his daughters

57. All of the following statements relating to gene mutations are valid EXCEPT:

(A) Gene mutations involve an alteration in the nitrogen-base sequence of DNA.
(B) Gene mutations are usually inherited as dominant alleles.
(C) Gene mutations may occur spontaneously, involving no apparent mutagenic agent.
(D) A change in only one nitrogen base may be sufficient to cause the production of a defective protein.
(E) Only those mutations occurring in germ cells can be passed along to future generations.

58. The goal of a teaching unit is for students to recognize that species are replaced by other species as their environment changes. Which of the following is an appropriate learning experience?

(A) Comparing plant types of a primary dune to those of a secondary dune
(B) Characterizing the stages of embryo development of a chick
(C) Observing phototropic movements by plants
(D) Recording changes in heart rate of *Daphnia* at different temperatures
(E) Comparing the brain structure and function among higher vertebrates

59. All of the following statements relating to the fossil record are correct EXCEPT:

(A) The most ancient fossils are situated in the oldest layers of rock.
(B) Aquatic organisms evolved before land organisms.
(C) Many species have become extinct over time.
(D) Organisms have diversified over time.
(E) Dinosaur and human remains are usually found within the same rock layer.

60. In many animal societies, a social hierarchy, or peck order, controls the dominance behavior of its members. This peck order generally serves to accomplish all the following EXCEPT

(A) reduce aggression among the members of the society
(B) establish a means by which the members of the society obtain mates
(C) influence the way in which the members of the society distribute themselves within the living space
(D) establish a means by which food is distributed among the members of the society
(E) provide a means whereby the weak members of the society are protected and cared for

GO ON TO THE NEXT PAGE.

61. An atmospheric relative humidity of 50 percent means that

(A) there is a 50 percent probability of rain
(B) half of the moisture in the air is condensing in the form of dew and droplets on objects
(C) the atmosphere contains 50 grams of water vapor per cubic meter
(D) the atmosphere is saturated with water vapor when the air temperature is 50° F
(E) the atmosphere contains half as much water vapor as it could contain at its present temperature

62. Similarities in the bone structure of the foreleg of a horse, the wing of a bat, and the forelimb of a whale are examples of

(A) inheritance from a common ancestor
(B) inheritance of acquired characteristics
(C) evolution by use and disuse
(D) geographic distribution of species
(E) convergent evolution

63. In the Antarctic Ocean, fish have been found that can withstand freezing temperatures because of their ability to produce antifreeze compounds in their tissues. The accepted explanation for this phenomenon is that the

(A) cold temperatures caused ancestors of the fish to produce the compounds
(B) materials needed to form the compounds are available only in cold oceans
(C) fish that did not produce the compounds migrated to other areas
(D) isolation prevented other types of fish from entering the area
(E) fish that produced the compounds survived and reproduced

64. Evidence from lakes in southeastern Canada and the northeastern United States indicates that lakes are becoming depleted of living organisms. Which of the following is considered the primary cause?

(A) Nuclear waste
(B) Lowered water levels
(C) Acid rain
(D) Garbage dumping
(E) Oil spills

65. The explosive growth and the success of rabbits in Australia after their introduction are due mainly to

(A) disruptive selection
(B) stabilizing selection
(C) directional selection
(D) reduced selection pressure
(E) isolation

66. Which of the following types of behavior is manifested by a male redwing blackbird who attacks any blackbird that approaches the area where he is displaying?

(A) Dominance
(B) Latent learning
(C) Conditioned reflex
(D) Appeasement
(E) Territoriality

67.

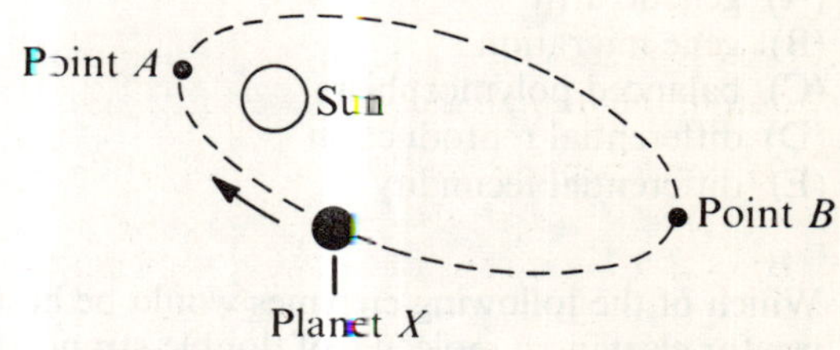

According to the laws of planetary motion, which of the statements below describes the motion of planet X as it moves in its orbit through points A and B in the diagram above?

(A) Its speed increases as it approaches point A and decreases as it approaches point B.
(B) Its speed decreases as it approaches point A and increases as it approaches point B
(C) Its speed remains unchanged as it approaches point A but increases as it approaches point B.
(D) Its speed remains unchanged as it approaches point A but decreases as it approaches point B.
(E) Its speed is constant through all points on its orbit.

GO ON TO THE NEXT PAGE.

68. Bioengineering through new genetic technology is revolutionizing modern medical testing procedures. One new method designed to test for the presence of a defective allele even when the exact nucleotide sequence change is not known is called

(A) karyotyping of humans
(B) ultrasound fetal monitoring
(C) restriction-fragment-length polymorphism (RFLP) testing
(D) amniocentesis
(E) chorionic villi sampling

69. It is known that heterozygous individuals living in central Africa who carry a gene for normal hemoglobin and a gene for sickle cell anemia are resistant to malaria. Whenever a heterozygous condition confers a selective survival advantage over homozygous types, this situation results in a phenomenon called

(A) genetic drift
(B) gene migration
(C) balanced polymorphism
(D) differential reproduction
(E) differential fecundity

70. Which of the following enzymes would be best to use for cleaving a molecule of double-stranded DNA into segments of various sizes?

(A) Restriction enzyme
(B) DNA polymerase
(C) Exonuclease
(D) Reverse transcriptase
(E) DNA ligase

71. Eggs with shells evolved first in which of the following groups?

(A) Fishes
(B) Amphibians
(C) Birds
(D) Mammals
(E) Reptiles

72. All of the atoms of sodium, whether or not in compounds, must have exactly eleven of which of the following?

(A) ions
(B) protons
(C) neutrons
(D) electrons
(E) photons

73. School policy statements concerning the handling of students who develop AIDS can be best used by classroom teachers to

(A) introduce a unit on bacteria
(B) teach a lesson on DNA
(C) develop concepts about laws pertaining to genetic screening
(D) show how scientific information relates to decisions made in society
(E) explain how a vaccine is made

74. During metaphase in the process of mitosis, the line-up of chromosomes along the equator helps ensure that

(A) the newly formed daughter cells contain a copy of each chromosome
(B) the homologous chromosomes stay within the nucleus of the cell
(C) the chromatid pairs remain together
(D) gametes are formed
(E) crossing-over occurs

75. The cellular synthesis of RNA from a DNA template is called

(A) duplication
(B) translation
(C) transcription
(D) transformation
(E) replication

76. A population curve shows a rapid increase initially, then levels off to a rather stable number of individuals. At this stable level, the population has reached

(A) the biotic potential
(B) the carrying capacity
(C) the saturation point
(D) a free-running period
(E) biological magnification

77. Which of the following local changes in atmospheric conditions would most likely result in rain?

(A) A decrease in relative humidity
(B) A decrease in atmospheric pressure
(C) An increase in carbon dioxide concentration
(D) An increase in air density
(E) A replacement of cool, humid air by warm, dry air

GO ON TO THE NEXT PAGE.

78. $\ldots C_8H_{18} + \ldots O_2 \rightarrow \ldots CO_2 + \ldots H_2O$

When the equation above is balanced with the lowest whole-number coefficients, what is the coefficient for CO_2?

(A) 4
(B) 8
(C) 12
(D) 16
(E) 18

79. If a series circuit powered by a 6-volt battery has a current of 2 amperes, the total resistance in that circuit is

(A) 3 ohms
(B) 4 ohms
(C) 6 ohms
(D) 8 ohms
(E) 12 ohms

80. Which of the following hypotheses is LEAST amenable to verification by experimentation?

(A) The average size of eukaryotic genes is greater than the average size of prokaryotic genes.
(B) *Archaeopteryx* evolved from a reptilian ancestor.
(C) Triploid watermelons exhibit defective meiosis and are therefore seedless.
(D) The DNA that codes for the beta chain of hemoglobin has more introns than the DNA that codes for the alpha chain.
(E) Humans and chimpanzees share some of the same genes.

81. FOOD FROM 2.27 KILOGRAMS OF GRAIN

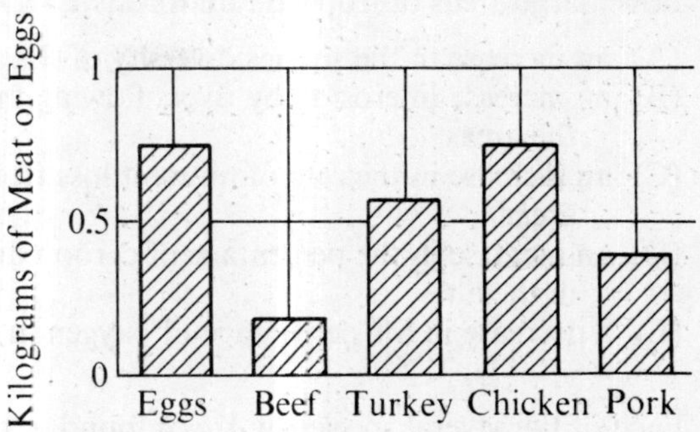

The graph above illustrates the numbers of kilograms of meat or eggs produced for every 2.27 kilograms of grain eaten by an animal. Which of the following changes in the eating habits of people in the developed world would result in the most efficient use of grain?

(A) Eating more pork than eggs
(B) Eating more chicken than beef
(C) Eating more beef than eggs
(D) Eating more turkey than chicken
(E) Eating more eggs than chicken

82. According to the theory of plate tectonics, which of the following statements is correct?

(A) Crustal plates are destroyed at the midocean ridges.
(B) The Earth's crust is one solid plate made of basalt.
(C) The outer core of the Earth is made of solid material.
(D) Earthquakes and volcanoes usually occur in the middle of crustal plates.
(E) Continents drift because they are located on crustal plates that are moving.

83. From the perspective of its evolutionary origins, which of the animal species below is most distantly related to humans?

(A) Bat
(B) Tree shrew
(C) Whale
(D) Opossum
(E) Raccoon

GO ON TO THE NEXT PAGE.

84. All of the following are likely to result from cutting down large areas of tropical rain forests EXCEPT

(A) an increase in the species diversity of the areas
(B) an increase in erosion by rivers flowing through the areas
(C) an increase in the rate of nutrient loss from the areas
(D) an increase in the percentage of carbon dioxide in the air
(E) a decrease in the percentage of oxygen in the air

85. Each of the several species of *Yucca* found in the North American desert can only be pollinated by a unique species of yucca moth. Each species of moth is dependent on its species of plant for food and reproduction. This relationship is best described as

(A) parasitism
(B) endosymbiosis
(C) mutualism
(D) commensalism
(E) predation

86. Which of the following elements makes up the greatest percentage by mass of the Earth's crust?

(A) Iron
(B) Carbon
(C) Nitrogen
(D) Silicon
(E) Sulfur

87. The presence of which of the following is useful in separating ferns and mosses into different taxonomic categories?

(A) Chlorophyll
(B) Vascular tissue
(C) Nuclei
(D) Chloroplasts
(E) Flagellated sperm

88. When moose migrate, the weaker ones often become the prey of wolves and other carnivores. If, in a particular year, vegetation eaten by moose was sparse in an area, which of the following statements is most likely correct concerning the wolf population?

(A) The wolf population will not change.
(B) The wolf population will initially increase due to an increase in food.
(C) The wolf population will initially decrease because moose would become stronger by fighting over the food.
(D) The wolf population will increase simultaneously with the moose population.
(E) The gain in wolf biomass will be equal to the quantity of the moose eaten.

89.

I. Fish skeleton
II. Rodent skull
III. Trilobite
IV. Cyanobacteria
V. Salamander-like creature

Students were presented with the representative fossils listed above. They were asked to place the fossils in the order in which their respective groups evolved. Starting with the oldest group, which of the following sequences is correct?

(A) I, II, III, IV, V
(B) II, V, I, III, IV
(C) III, IV, II, V, I
(D) IV, III, I, V, II
(E) IV, III, V, I, II

90. A student presents a current-events report on using engineered microorganisms to clean up specific hazardous-waste sites. Of the following questions posed by other students, which has the LEAST direct relevance to the topic of the report?

(A) How can microorganisms be engineered?
(B) Why can these microorganisms digest the wastes and other microorganisms cannot?
(C) What happens to the wastes?
(D) How do microorganisms cause disease in humans?
(E) How can these microorganisms fit into the scheme of other decomposers in the area?

91. Which of the following isolating mechanisms is generally considered to initiate the process of speciation in the majority of cases?

(A) Mechanical isolation
(B) Behavioral isolation
(C) Developmental isolation
(D) Seasonal isolation
(E) Geographic isolation

92. A segment of DNA that codes for a polypeptide containing 120 amino acids must have a minimum of how many nucleotide base pairs?

(A) 40
(B) 120
(C) 240
(D) 360
(E) 480

GO ON TO THE NEXT PAGE.

93. On a lush 35-square-mile island, fifty miles from the nearest coast of the mainland, 5 male deer and 17 female deer were introduced. No predators of deer were on the island and no hunting was allowed. The graph below shows the population of deer over 50 years.

DEER POPULATION GRAPH

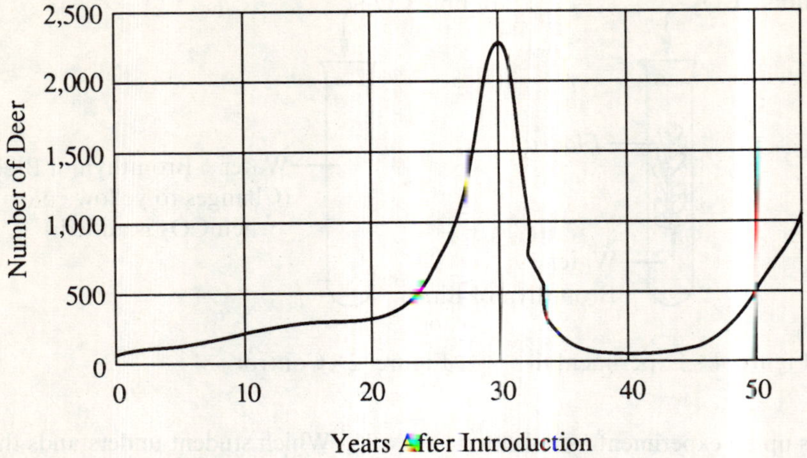

Statements that are probably correct about the deer population include which of the following?

I. The deer multiplied because of the abundance of food.
II. The deer reached high population density because of a lack of predators.
III. The deer population could have increased indefinitely if it were not for diseases that probably spread through the herd.

(A) I only
(B) II only
(C) III only
(D) I and II only
(E) II and III only

---

94. Several plant species growing on moraines of lower New York State and New Jersey are known to grow otherwise only in northern Canada. Which of the following mechanisms is most likely responsible for the existence of these plant species in the United States?

(A) Migration of foxes
(B) Glaciation
(C) Divergent evolution
(D) Shifts in magnetic poles
(E) Rising levels of global oceans

95. Which of the following can be found in both prokaryotic cells and eukaryotic plant cells?

(A) Mitochondria
(B) Chloroplasts
(C) Ribosomes
(D) Endoplasmic reticulum
(E) Golgi complex

GO ON TO THE NEXT PAGE.

Directions: The groups of questions below are based on classroom situations. First read the description and the students' comments. Then identify the appropriate student for each question and indicate your answer by filling in the corresponding space on the answer sheet. One comment may be used once, more than once, or not at all.

Questions 96-98

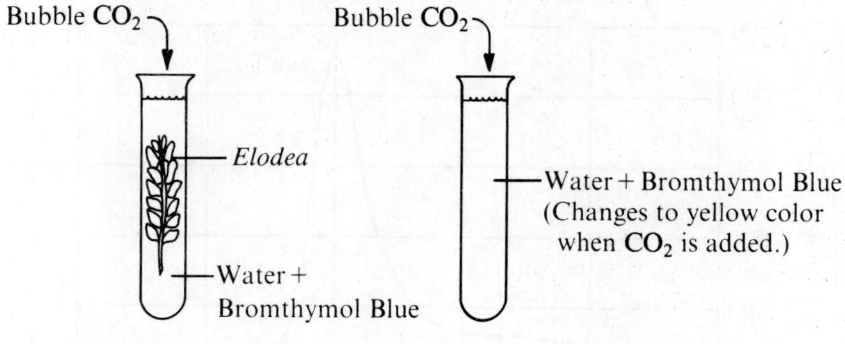

Figure 1 - Experimental   Figure 2 - Control

A biology teacher sets up an experiment as indicated in the diagrams above. A sprig of *Elodea*, water, and bromthymol blue indicator are placed in a test tube. Carbon dioxide is bubbled through the solution for a few minutes, which changes the color from blue to yellow. The plant is allowed to photosynthesize. During the day, the yellow solution turns blue; at night, it turns back to yellow. The control for this demonstration is a test tube of water and bromthymol blue, which turns yellow when carbon dioxide is bubbled through the solution.

The students are asked to explain what caused the color changes observed in the experimental tube shown in Figure 1. Five students respond as follows.

Student A: Bromthymol blue turns yellow in an acid medium and blue in a basic medium.

Student B: Photosynthesis uses up carbon dioxide, causing the solution to turn yellow, and cellular respiration causes the solution to turn blue.

Student C: Carbon dioxide, water, sunlight, and chlorophyll are all necessary for photosynthesis.

Student D: Oxygen is released during photosynthesis, which causes the solution to turn blue, then yellow at night.

Student E: During the day, photosynthesis uses CO_2 and causes the yellow color to change to blue; at night, the release of CO_2 by cellular respiration causes the blue color to change to yellow.

96. Which student understands the requirements of photosynthesis, but has not answered the question?

97. Which student best answers the question?

98. Which student best understands the chemical reasons for the color change, but did not answer the question?

GO ON TO THE NEXT PAGE.

Questions 99-102

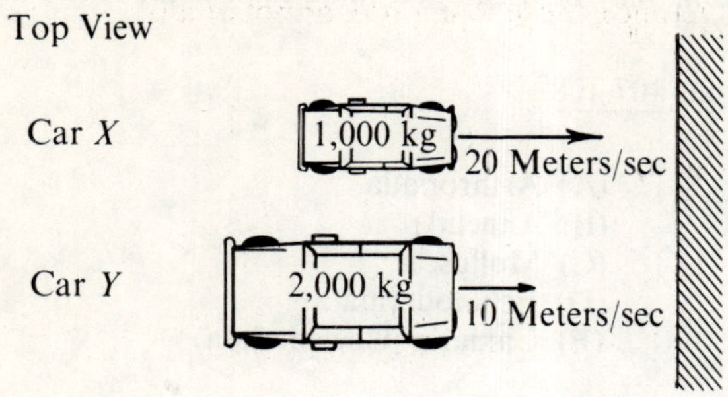

During a discussion of momentum and kinetic energy, the teacher draws on the board the diagram above, and asks the class to compare the momentums and the kinetic energies of the two cars. Five students respond as follows:

Student A: Car *X* is lighter and so both its momentum and its kinetic energy are less than the values for car *Y* are.

Student B: There is too little information for me to make any comparisons.

Student C: The kinetic energies are the same, but car *Y* has greater momentum since it is heavier.

Student D: Since both momentum and kinetic energy include both a mass term and a speed term, and since car *X* has twice the speed of car *Y* but car *Y* weighs twice as much as car *X*, the kinetic energy and the momentum for the two cars are the same.

Student E: The momentums are the same, but car *X* has the greater kinetic energy since its speed is greater and kinetic energy has a speed squared term.

99. Which student has the best answer to the teacher's question?

100. Which student has demonstrated the LEAST understanding of the concepts involved?

101. Which student has shown enough understanding that seeing the equation defining kinetic energy should lead the student to correct the error?

102. Which student has given an answer that would be correct if the two cars were moving at the same speeds?

GO ON TO THE NEXT PAGE.

Directions: Each group of questions below consists of five lettered headings followed by a list of numbered words, phrases, or sentences. For each numbered word, phrase, or sentence, select the one heading that is most closely related to it and fill in the corresponding space on the answer sheet. One heading may be used once, more than once, or not at all in each group.

Questions 103-104

(A) Leech
(B) Mosquito
(C) Lichen
(D) Fungus
(E) Flea

103. Is a parasitic annelid

104. Is an example of a mutualistic relationship

Questions 105-106

(A) Anton van Leeuwenhoek
(B) Linus Pauling
(C) Louis Pasteur
(D) Stanley Miller
(E) Carolus Linnaeus

105. Provided the system of binomial nomenclature currently used to classify plants and animals

106. Performed an experiment that demonstrated amino acid synthesis in a simulated primitive atmosphere

Questions 107-108

(A) Arthropoda
(B) Annelida
(C) Mollusca
(D) Echinodermata
(E) Cnidaria (Coelenterata)

107. Hydra, jellyfish, sea anemone

108. Lobster, spider, tick

Questions 109-111

(A) Protein
(B) Carbohydrate
(C) Alcohol
(D) Nucleic acid
(E) Steroid

109. Cholesterol

110. DNA

111. Glycogen

GO ON TO THE NEXT PAGE.

Questions 112-115

(A) Infrared rays
(B) Ultraviolet rays
(C) Radio waves
(D) Visible rays
(E) X-rays

112. Are detected by large antenna arrays and provide detailed images of astronomical objects such as quasars

113. Are associated with vibrations in molecules and generally perceived as heat

114. Primarily account for the perception of color as a result of selective absorption or reflection

115. Are associated with beta-particle emission by radioactive nuclides and with ejection of atomic K-shell electrons

Questions 116-118

(A) Osmosis
(B) Transpiration
(C) Pinocytosis
(D) Phagocytosis
(E) Translocation

116. The intake of fluid droplets by a cell

117. The diffusion of water through a semipermeable membrane from a region of low solute concentration to a region of high solute concentration

118. The taking of solid materials into a cell by engulfment and the subsequent pinching off of the cell membrane

GO ON TO THE NEXT PAGE.

Questions 119-121 refer to the diagram below of the cross section of a leaf.

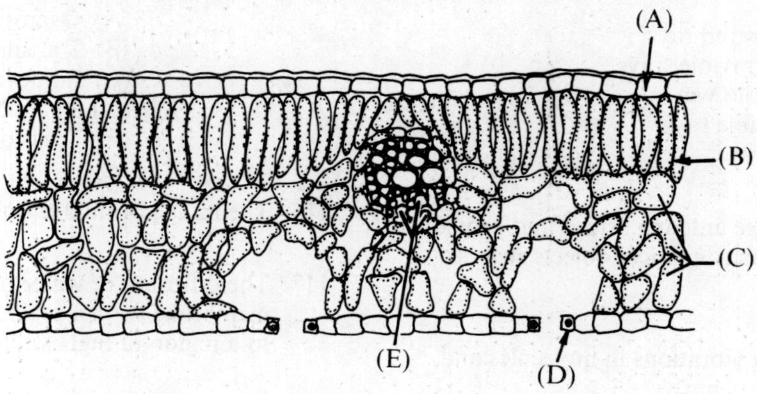

119. A site of greatest photosynthetic activity

120. A site of translocation

121. A site that regulates the amount of water lost from the leaf

---

Questions 122-125

(A) Complex ion formation
(B) Single replacement
(C) Double replacement (exchange of ions)
(D) Direct combination (synthesis)
(E) Decomposition (analysis)

122. $CaO + H_2O \rightarrow Ca(OH)_2$

123. $2\ NaCl \rightarrow 2\ Na + Cl_2$

124. $Cu + 2\ AgNO_3 \rightarrow 2\ Ag + Cu(NO_3)_2$

125. $AgNO_3 + NH_4Cl \rightarrow AgCl + NH_4NO_3$

GO ON TO THE NEXT PAGE.

Directions: Each question below consists of an incomplete analogy followed by five suggested completions. Each analogy, when properly completed, would be suitable for use in explaining the scientific terms in the incomplete analogy. Select the pair of terms that best completes the analogy and indicate your selection by filling in the corresponding space on the answer sheet.

126. IGNEOUS : ROCK ::
(A) potential : energy
(B) metamorphic : magma
(C) measured : liter
(D) infrared : radioactivity
(E) sedimentary : granite

127. ENDOTHERMIC : EXOTHERMIC ::
(A) acid : base
(B) refrigerator : furnace
(C) window : door
(D) cathode : anode
(E) proton : electron

128. WIND SPEED : AIR PRESSURE GRADIENT ::
(A) erosion : deposition of sediment
(B) relative humidity : dew point
(C) river velocity : slope of riverbed
(D) air temperature : altitude
(E) heat : temperature

129. ANODE : CATHODE ::
(A) rate : speed
(B) temperature change : volume change
(C) saturated solution : supersaturated solution
(D) pressure : contraction
(E) oxidation : reduction

130. SCIENTIFIC METHOD : TRIAL AND ERROR ::
(A) pendulum : clock
(B) lever : fulcrum
(C) pump : handle
(D) professional : novice
(E) doctor : patient

131. ADENINE : THYMINE ::
(A) adenine : guanine
(B) adenine : cytosine
(C) cytosine : uracil
(D) cytosine : thymine
(E) guanine : cytosine

132. VOYAGER : NEPTUNE ::
(A) oscilloscope : light
(B) electron microscope : mitochondria
(C) Bunsen burner : heat
(D) filter : water
(E) motor : boat

GO ON TO THE NEXT PAGE.

Directions: Each group of questions below concerns a laboratory or experimental situation. In each case, first study the description of the situation. Then choose the one best answer to each question following it and fill in the corresponding space on the answer sheet.

Questions 133-137

The figure below shows a food web. Each letter represents a species. The arrows show the flow of materials and energy.

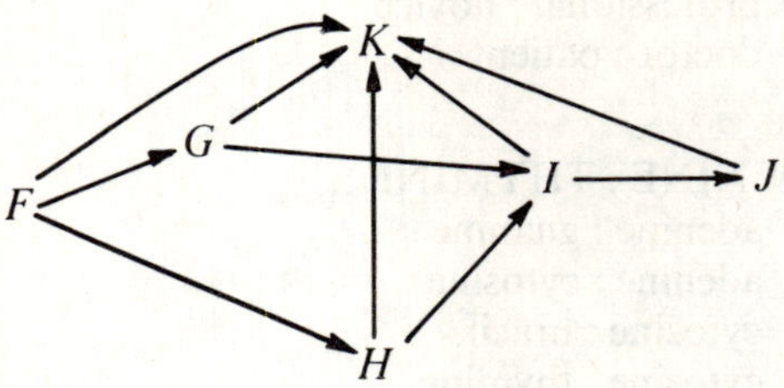

133. The producer in this food web is

(A) *F*
(B) *H*
(C) *I*
(D) *J*
(E) *K*

134. Which of the following would have the greatest total biomass?

(A) *F*
(B) *J*
(C) *K* alone
(D) *G* + *H*
(E) *I* + *K*

135. *J* is a

(A) primary consumer
(B) secondary consumer
(C) tertiary consumer
(D) decomposer
(E) primary producer

136. Suppose that a toxic substance becomes concentrated in the food web by biological magnification. Of the following, the species likely to be most affected is

(A) *F*
(B) *G*
(C) *H*
(D) *I*
(E) *J*

137. Which of the following species is most likely the decomposer?

(A) *F*
(B) *H*
(C) *I*
(D) *J*
(E) *K*

Questions 138-140

In a certain species of mammal, the allele for brown fur is dominant over the allele for white fur and the allele for long fur is dominant over that for short fur. An animal heterozygous for long, brown fur is mated to an animal that is homozygous for short, white fur. Assume that these alleles assort independently.

138. The percentage of the F_1 offspring expected to have white fur is

(A) 100%
(B) 75%
(C) 50%
(D) 25%
(E) 0%

139. The percentage of the F_1 offspring expected to be heterozygous for both fur color and fur length is

(A) 100%
(B) 75%
(C) 50%
(D) 25%
(E) 0%

140. The percentage of the F_1 offspring expected to be homozygous for brown fur is

(A) 100%
(B) 75%
(C) 50%
(D) 25%
(E) 0%

GO ON TO THE NEXT PAGE.

<u>Questions 141-143</u>

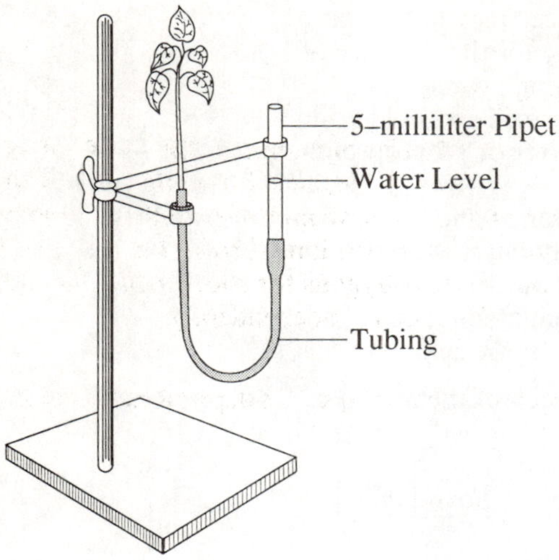

The stem of a plant shoot was held beneath water and cut diagonally. Tubing attached at one end to a pipet was filled with water and the tubing forced over the cut end of the shoot while still immersed in the water. The tubing and pipet were clamped as shown in the diagram above. Measurements were taken by reading the water level in the pipet every 15 minutes.

141. The experimental setup is designed to measure

(A) root pressure
(B) photosynthesis
(C) transpiration
(D) guttation
(E) translocation

142. Which of the following would have the LEAST effect on changing the water level from its initial value?

(A) Ambient temperature
(B) Number of leaves
(C) Humidity
(D) Dissolved CO_2 in the water
(E) Number of open stomata

143. Which of the following actions is most likely to lead to the most rapid lowering of the water level in the pipet?

(A) Increasing the light intensity
(B) Sprinkling the leaves with water
(C) Covering the plant with a clear plastic bag
(D) Covering the leaves with petroleum jelly
(E) Covering the plant with a green plastic bag

GO ON TO THE NEXT PAGE.

Questions 144-147

Catalase is an enzyme that facilitates the breakdown of hydrogen peroxide, a harmful product of cell metabolism, into water and oxygen. In an experiment to test for the presence of catalase in tissue, students used the following procedures.

Ten milliliters of 3 percent hydrogen peroxide was placed in a series of numbered tubes and pieces of liver or potato were dropped in using forceps.

The ground tissue was prepared in a mortar and pestle with some sand as an abrasive. The cooked material was prepared by boiling it in water in a test tube over a Bunsen burner. Liver was soaked for 24 hours in 70 percent ethanol or 0.1-molar lead nitrate or 0.1-molar hydrochloric acid.

The following data were collected for each of the materials tested.

Material	Foaming
Sand	−
Whole liver	+
Ground liver	++
Cooked liver	−
Whole potato	+
Ground potato	++
Cooked potato	−
Liver soaked in 70% ethanol	−
Liver soaked in 0.1-M lead nitrate solution	−
Liver soaked in 0.1-M hydrochloric acid	−

Key: (−) no foaming (+) foaming (++) much foaming

144. Ground tissue produces more foaming than whole tissue because

(A) lipids from the disrupted cell membranes absorb more hydrogen peroxide
(B) water from the disrupted vacuoles accelerates the chemical process
(C) more enzyme is released from the disrupted cells to react with the substrate
(D) lysosomes from the ground tissue are disrupted and release proteolytic enzymes
(E) mitochondria from the disrupted cells have greater access to molecular oxygen

145. Which of the following methods is most convenient to test for the production of oxygen in this experiment?

(A) PTC paper
(B) Litmus paper
(C) Lime water
(D) Chromatography
(E) A glowing splint

146. Which of the following is the best explanation for the fact that the cooked tissue produced no foaming?

(A) The amino acid sequence in catalase has been changed.
(B) The cell membrane thickens with heating and prevents the release of enzymes.
(C) The messenger RNA has been destroyed and new proteins cannot be synthesized.
(D) The tertiary structure of catalase has been altered.
(E) The water in the cells has been evaporated and no chemical reaction can occur.

147. Based on the recorded data, which of the following experimental variables seems to have the LEAST effect on enzyme function?

(A) High temperature
(B) Lead nitrate
(C) Alcohol
(D) Tissue type
(E) Hydrochloric acid

GO ON TO THE NEXT PAGE.

Questions 148-151

A student in a laboratory is given 7 grams of iron filings and 4 grams of powdered sulfur. The two elements are thoroughly mixed together on a sheet of paper. (Molecular weights: iron = 55.85, sulfur = 32.07)

148. Which of the following is the best method for separating the iron from the mixture?

(A) Lifting out the iron particles by using a magnifying glass and tweezers
(B) Adding water and filtering
(C) Adding water and evaporating to dryness
(D) Heating the mixture
(E) Passing a magnet just above the mixture

149. If the mixture of iron and sulfur is heated in a test tube, a red glow spreads throughout the tube. Heat energy is obviously released. After cooling, the test tube and its contents are weighed. If the empty test tube weighed 17 grams and if it is assumed that no atmospheric gases were involved in the reaction, the mass of the test tube and its contents should be approximately

(A) 6 grams
(B) 17 grams
(C) 27 grams
(D) 28 grams
(E) 29 grams

150. If the test tube in which the iron and sulfur were heated is broken and the contents are mixed with carbon disulfide, none of the solid dissolves. Free sulfur is soluble in carbon disulfide. Conclusions supported by the information given include which of the following?

I. A chemical change has occurred.
II. The sulfur has completely reacted.
III. Iron is not soluble in carbon disulfide.

(A) I only
(B) I and II only
(C) I and III only
(D) II and III only
(E) I, II, and III

151. The most probable name of the product formed by the reaction of iron filings with powdered sulfur is

(A) iron III (ferric) sulfate
(B) iron III (ferric) sulfite
(C) iron II (ferrous) sulfite
(D) iron II (ferrous) sulfide
(E) iron II (ferrous) oxide

GO ON TO THE NEXT PAGE.

Questions 152-155 refer to the graph below, which depicts the oxygen-dissociation curves of the hemoglobin of various organisms.

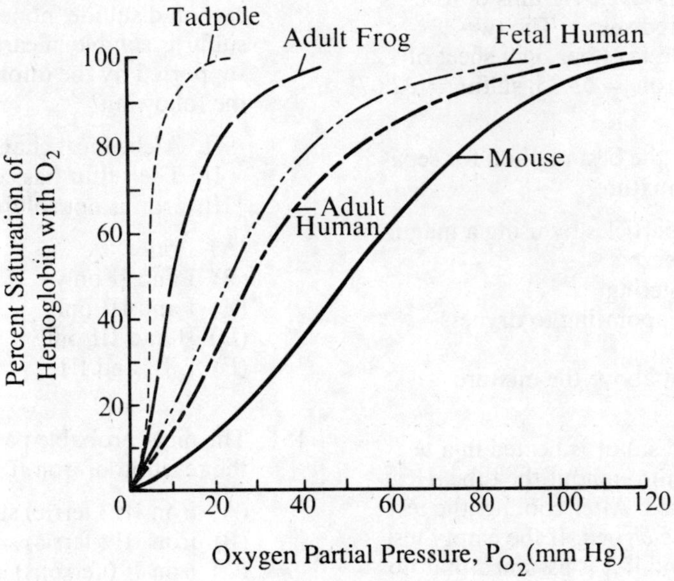

152. If the P_{O_2} of mouse blood is 60 millimeters of mercury, what percentage of hemoglobin is saturated with oxygen?

(A) 10%
(B) 30%
(C) 60%
(D) 80%
(E) 100%

153. Which of the animals shows the greatest rate of change in oxygen uptake between 80 and 120 millimeters of mercury?

(A) Tadpole
(B) Adult frog
(C) Fetal human
(D) Adult human
(E) Mouse

154. The oxygen-dissociation curves provide evidence that

(A) tadpole hemoglobin has a greater affinity for oxygen than has the hemoglobin of an adult frog
(B) at all partial pressures of oxygen, fetal human hemoglobin has the same affinity for oxygen as has the hemoglobin of the adult human
(C) the carrying capacity of hemoglobin for oxygen is less in younger animals than in older animals
(D) a tadpole lives in a more oxygen-rich environment than an adult frog does
(E) a fetal human lives in a more oxygen-rich environment than a mouse does

155. Of the following, the last organism to succumb in an oxygen-deprived environment would probably be the

(A) tadpole
(B) adult frog
(C) fetal human
(D) adult human
(E) mouse

GO ON TO THE NEXT PAGE.

Questions 156-160

The ability of tobacco plants to produce chlorophyll is governed by a dominant allele, *C*. The homozygous recessive condition, *cc*, produces albino tobacco plants.

In the laboratory, a student places 100 tobacco seeds on moist filter paper in each of 2 petri dishes. Both dishes are placed at room temperature for 3 days to allow seeds to germinate. Dish A is placed in the light. Dish B is placed in the dark. The student then records the data in Table 1 below.

TABLE 1

	Condition	Number of Non-Green Plants	Number of Green Plants
Dish A	Light	24	73
Dish B	Dark	96	0

Both dishes are left in the light for another day. The following data are then recorded.

TABLE 2

	Condition	Number of Non-Green Plants	Number of Green Plants
Dish A	Light	24	73
Dish B	Light	26	70

156. The variable that was most important in causing the results shown in Table 1 is the

(A) temperature
(B) genetic differences between the seeds in dish A and dish B
(C) amount of moisture
(D) presence of light
(E) lack of soil for germination

157. The genotype of the non-green seedlings in dish A is most likely

(A) *cc* only
(B) *Cc* only
(C) *CC* only
(D) *Cc* or *cc*
(E) *CC* or *cc*

158. The green seedlings in dish A would most likely have which of the following genotypes?

(A) *cc* only
(B) *Cc* only
(C) *CC* only
(D) *CC* and *cc*
(E) *CC* and *Cc*

159. From the results obtained in dish B, it can be correctly concluded that

(A) tobacco-seed genes spontaneously mutate under certain environmental conditions
(B) environmental factors influence gene expression
(C) gene expression is too unpredictable to be of any scientific value
(D) color in tobacco seedlings is not influenced by genes
(E) chlorophyll promotes the germination of seedlings

160. From the results obtained in Table 2, the genotypes of the parent plants of the seedlings are most likely to be which of the following?

(A) *CC* × *CC*
(B) *CC* × *Cc*
(C) *CC* × *cc*
(D) *Cc* × *Cc*
(E) *cc* × *cc*

S T O P

IF YOU FINISH BEFORE TIME IS CALLED, YOU MAY CHECK YOUR WORK ON THIS TEST.

NTE Programs

BIOLOGY AND GENERAL SCIENCE

The supervisor will tell you when to begin work on the test and when to stop. If you finish the test before time is called, go back and check your work on it.

SHOULD YOU GUESS? In this test your score is based on the number of questions you answer correctly; therefore skipped and wrong answers will not count against you. Work as rapidly as you can without sacrificing accuracy. Do not spend too much time puzzling over a question that seems too difficult for you. Answer the easier questions first, then return to the harder ones. Try to answer every question even if you have to guess.

Where necessary, you may use blank spaces in the test book for scratch paper. Do not use any other paper or the margins or back of the answer sheet to do scratchwork.

YOU ARE TO INDICATE ALL OF YOUR ANSWERS ON THE SEPARATE ANSWER SHEET. No credit will be given for anything written in this examination book. After you have decided which of the suggested answers is best, fill in the corresponding space on the answer sheet. BE SURE THAT EACH MARK IS HEAVY AND DARK AND COMPLETELY FILLS THE ANSWER SPACE. Light or partial marks may not be read by the scoring machine. Give only one answer to each question. If you change an answer, be sure that the previous mark is erased completely. Incomplete erasures may be read as intended answers.

This test may include one or more questions that do not count toward your score.

Time—120 minutes

160 Questions

DO NOT BREAK THE SEAL UNTIL YOU ARE TOLD TO DO SO.

NTE® PROGRAMS SPECIALTY AREA TEST

USE ONLY A SOFT LEAD PENCIL FOR COMPLETING THIS ANSWER SHEET. ERASE NEATLY AND COMPLETELY.

SIDE 1

1. NAME Omit spaces, hyphens, apostrophes, and Jr., III, etc.

Last (family) Name-first 12 letters | First (given) Name-first 8 letters | M.I.

2. SEX

Male ○

Female ○

3. YOUR MAILING ADDRESS • Fit address into boxes provided. • Indicate a space in address by leaving a blank box and filling in the corresponding diamond.

P.O. Box or Street Address | City | State or Province

3a. Zip/ Postal Code

Please enter code and fill in the corresponding ovals.

BACKGROUND INFORMATION—Answers to questions 4a and 4b will be used for research purposes only. In no case will an individual's responses be singled out.

4a. If you have a choice, which type of test **in this subject** would you prefer to take?

○ Multiple-choice ○ Open-ended: essay, problems, etc. ○ No preference

4b. Do you understand English as well as or better than any other language?

○ Yes ○ No

5. How do you describe yourself?

Fill in only one oval.

- ○ Afro American or Black (non-Hispanic)
- ○ Mexican American or Chicano
- ○ Native American, American Indian, or Alaskan Native
- ○ Pacific/Asian American
- ○ Puerto Rican
- ○ Other Hispanic or Latin American
- ○ White (non-Hispanic)
- ○ Other

6. What is your cumulative undergraduate average to date (based on a system where 4.0 = A)?

- ○ 3.5 – 4.0
- ○ 3.0 – 3.49
- ○ 2.5 – 2.99
- ○ 2.0 – 2.49
- ○ 1.5 – 1.99
- ○ below 1.5

7. Major Field Codes

Use your Critical Information Form to complete this item. (If you do not have a Critical Information Form, leave this item blank.)

MAJOR FIELD CODES

UNDERGRADUATE | GRADUATE

8. DATE OF BIRTH

MONTH | DAY | YEAR

9. SOCIAL SECURITY NUMBER (optional)

10. EDUCATIONAL LEVEL

What is the highest educational level you have reached?

Fill in only one oval.

- ○ Freshman
- ○ Sophomore
- ○ Junior
- ○ Senior
- ○ Hold bachelor's degree
- ○ Enrolled in graduate school
- ○ Hold master's degree
- ○ Hold doctoral degree

I.N.274244 51055 • 01380 • TF51P250 MH91142

NTE® PROGRAMS SPECIALTY AREA TEST

SIDE 2

11. REGISTRATION NUMBER

12. TEST CENTER NUMBER AND LOCATION

NUMBER

PLEASE PRINT:

Center Name

City

State

13. TODAY'S DATE

MO / DAY / YR

ETS USE ONLY

Use your completed Critical Information Form for items 14, 15, and 16. If you do not have a code number for one or more of these items, leave the item(s) blank. Scores will be sent only to agencies that are authorized score recipients.

14. COLLEGE WHERE YOU RECEIVED TRAINING RELEVANT TO THE TEST

CODE

ETS USE ONLY

15. AGENCIES TO RECEIVE SCORE REPORTS

R NUMBER 1 CODE

R NUMBER 2 CODE

R NUMBER 3 CODE

16. STATE/AGENCY CODE

CODE

17. YOUR MAILING ADDRESS PRINT COMPLETE ADDRESS BELOW.

Number and Street

Apt. (if any)

City

State/Province

Zip/Postal Code

Telephone Number: ( )

18. SIGNATURE AND STATEMENT

Please write the following statement in longhand on the lines below.

"I am the person whose name and address appear on this answer sheet."

All NTE tests are confidential. I agree not to disclose to anyone the contents of any NTE test I take, and I accept the conditions set forth in the *Bulletin of Information* concerning the administration of the tests and the reporting of scores.

Sign your full first name, middle name or initial, and last name.

19. TEST BOOK INFORMATION

SERIAL NUMBER

TEST FORM

TEST NAME

20. TEST CODE

BE SURE EACH MARK COMPLETELY FILLS THE INTENDED SPACE AND DOES NOT EXTEND INTO THE ADJACENT OVALS. DO NOT MAKE ANY STRAY MARKS. MAKE ALL ERASURES COMPLETE.

1 A B C D E	26 A B C D E	51 A B C D E	76 A B C D E	101 A B C D E	126 A B C D E	151 A B C D E	176 A B C D E
2 A B C D E	27 A B C D E	52 A B C D E	77 A B C D E	102 A B C D E	127 A B C D E	152 A B C D E	177 A B C D E
3 A B C D E	28 A B C D E	53 A B C D E	78 A B C D E	103 A B C D E	128 A B C D E	153 A B C D E	178 A B C D E
4 A B C D E	29 A B C D E	54 A B C D E	79 A B C D E	104 A B C D E	129 A B C D E	154 A B C D E	179 A B C D E
5 A B C D E	30 A B C D E	55 A B C D E	80 A B C D E	105 A B C D E	130 A B C D E	155 A B C D E	180 A B C D E
6 A B C D E	31 A B C D E	56 A B C D E	81 A B C D E	106 A B C D E	131 A B C D E	156 A B C D E	181 A B C D E
7 A B C D E	32 A B C D E	57 A B C D E	82 A B C D E	107 A B C D E	132 A B C D E	157 A B C D E	182 A B C D E
8 A B C D E	33 A B C D E	58 A B C D E	83 A B C D E	108 A B C D E	133 A B C D E	158 A B C D E	183 A B C D E
9 A B C D E	34 A B C D E	59 A B C D E	84 A B C D E	109 A B C D E	134 A B C D E	159 A B C D E	184 A B C D E
10 A B C D E	35 A B C D E	60 A B C D E	85 A B C D E	110 A B C D E	135 A B C D E	160 A B C D E	185 A B C D E
11 A B C D E	36 A B C D E	61 A B C D E	86 A B C D E	111 A B C D E	136 A B C D E	161 A B C D E	186 A B C D E
12 A B C D E	37 A B C D E	62 A B C D E	87 A B C D E	112 A B C D E	137 A B C D E	162 A B C D E	187 A B C D E
13 A B C D E	38 A B C D E	63 A B C D E	88 A B C D E	113 A B C D E	138 A B C D E	163 A B C D E	188 A B C D E
14 A B C D E	39 A B C D E	64 A B C D E	89 A B C D E	114 A B C D E	139 A B C D E	164 A B C D E	189 A B C D E
15 A B C D E	40 A B C D E	65 A B C D E	90 A B C D E	115 A B C D E	140 A B C D E	165 A B C D E	190 A B C D E
16 A B C D E	41 A B C D E	66 A B C D E	91 A B C D E	116 A B C D E	141 A B C D E	166 A B C D E	191 A B C D E
17 A B C D E	42 A B C D E	67 A B C D E	92 A B C D E	117 A B C D E	142 A B C D E	167 A B C D E	192 A B C D E
18 A B C D E	43 A B C D E	68 A B C D E	93 A B C D E	118 A B C D E	143 A B C D E	168 A B C D E	193 A B C D E
19 A B C D E	44 A B C D E	69 A B C D E	94 A B C D E	119 A B C D E	144 A B C D E	169 A B C D E	194 A B C D E
20 A B C D E	45 A B C D E	70 A B C D E	95 A B C D E	120 A B C D E	145 A B C D E	170 A B C D E	195 A B C D E
21 A B C D E	46 A B C D E	71 A B C D E	96 A B C D E	121 A B C D E	146 A B C D E	171 A B C D E	196 A B C D E
22 A B C D E	47 A B C D E	72 A B C D E	97 A B C D E	122 A B C D E	147 A B C D E	172 A B C D E	197 A B C D E
23 A B C D E	48 A B C D E	73 A B C D E	98 A B C D E	123 A B C D E	148 A B C D E	173 A B C D E	198 A B C D E
24 A B C D E	49 A B C D E	74 A B C D E	99 A B C D E	124 A B C D E	149 A B C D E	174 A B C D E	199 A B C D E
25 A B C D E	50 A B C D E	75 A B C D E	100 A B C D E	125 A B C D E	150 A B C D E	175 A B C D E	200 A B C D E

FOR ETS USE ONLY | R | CS

PART FOUR: SCORING THE PRACTICE TEST

How to Score the Biology and General Science Practice Test

Compare your answers with the correct answers in Table 1 below. Count the number of questions you answered correctly. Use Table 2 on page 44 to find the scaled score corresponding to the number answered correctly.

Table 1: Biology and General Science Test

Answers to Practice Test Questions and Percentages of Examinees Choosing Correct Answer

Question Number	Correct Answer	Content Category	Percent of Examinees Choosing Correct Answer	Question Number	Correct Answer	Content Category	Percent of Examinees Choosing Correct Answer	Question Number	Correct Answer	Content Category	Percent of Examinees Choosing Correct Answer
1	B	III	95.9	56	E	II	56.2	111	B	II	73.6
2	D	VII	29.9	57	B	II	56.8	112	C	VII	73.4
3	B	IV	74.5	58	A	V	84.3	113	A	VII	68.9
4	C	VII	91.6	59	E	IV	94.1	114	D	VII	80.1
5	E	VIII	72.3	60	E	V	67.8	115	E	VII	82.0
6	C	VII	76.3	61	E	VII	76.1	116	C	II	51.3
7	A	I	75.4	62	A	IV	54.6	117	A	II	92.7
8	D	VII	67.2	63	E	IV	87.6	118	D	II	84.4
9	C	VIII	33.5	64	C	I	81.1	119	B	III	45.0
10	B	IV	82.1	65	D	IV	59.0	120	E	III	51.8
11	A	V	68.6	66	E	V	93.5	121	D	III	68.5
12	B	VI	82.5	67	A	VIII	53.9	122	D	VI	70.4
13	E	I	94.0	68	C	I	41.2	123	E	VI	74.5
14	E	II	40.1	69	C	IV	34.5	124	B	VI	65.6
15	C	IV	39.6	70	A	I	17.5	125	C	VI	76.4
16	A	III	85.0	71	E	IV	62.5	126	A	VIII	62.0
17	C	V	92.4	72	B	VI	63.6	127	B	VI	51.6
18	A	V	47.4	73	D	[illegible]	91.2	128	C	VIII	47.4
19	B	IV	72.1	74	A	I	68.9	129	E	VI	75.9
20	D	III	70.1	75	C	I	58.3	130	D	I	65.3
21	D	VIII	79.6	76	B	V	64.7	131	E	II	78.9
22	E	IV	72.7	77	B	VII	58.2	132	B	VIII	77.4
23	C	V	87.5	78	D	VI	39.6	133	A	V	81.9
24	C	VI	81.8	79	A	VII	57.0	134	A	V	32.8
25	B	I	69.1	80	B	[illegible]	78.8	135	C	V	62.3
26	B	IV	18.0	81	B	[illegible]	76.0	136	E	V	52.9
27	B	VIII	84.8	82	E	VII	82.9	137	E	V	54.0
28	C	I	57.8	83	D	IV	24.6	138	C	III	56.6
29	B	IV	62.1	84	A	[illegible]	85.3	139	D	III	37.6
30	A	VIII	77.7	85	C	V	73.6	140	E	III	55.8
31	A	II	23.7	86	D	VIII	42.2	141	C	III	61.6
32	C	V	81.1	87	B	IV	67.6	142	D	III	56.7
33	E	VI	70.0	88	B	V	71.1	143	A	III	73.5
34	E	VI	87.5	89	D	IV	65.8	144	C	II	67.7
35	C	I	39.9	90	D	I	88.9	145	E	II	46.1
36	A	VIII	58.9	91	E	IV	78.8	146	D	II	39.2
37	D	III	48.5	92	D	I	30.0	147	D	II	63.5
38	E	I	71.4	93	D	V	66.7	148	E	VI	77.2
39	D	V	83.4	94	B	VIII	56.9	149	D	VI	59.1
40	C	VII	43.6	95	C	I	25.6	150	B	VI	34.4
41	E	VIII	46.8	96	C	II	82.3	151	D	VI	49.9
42	C	IV	52.6	97	E	II	77.3	152	C	III	85.1
43	D	V	70.3	98	A	II	59.4	153	E	III	63.8
44	C	III	39.6	99	E	VII	27.3	154	A	III	51.4
45	A	III	31.1	100	B	VII	82.5	155	A	III	59.8
46	A	V	85.4	101	D	VII	33.0	156	D	III	86.5
47	B	V	85.3	102	A	VII	34.0	157	A	III	71.0
48	D	III	48.3	103	A	II	82.9	158	E	III	69.3
49	A	II	61.8	104	C	II	77.7	159	B	III	73.6
50	E	VII	83.6	105	E	I	86.1	160	D	III	62.9
51	D	VII	54.8	106	D	I	54.7				
52	C	II	85.9	107	E	III	74.5				
53	C	III	34.3	108	A	III	83.2				
54	A	III	55.4	109	E	II	46.5				
55	D	III	83.3	110	D	II	87.3				

Note: The percentages in Table 1 are based on the test records of 1,140 examinees who took the Biology and General Science test at the March 1991 administration. The mean score for this group of examinees is BGS647 and the standard deviation of the scores is BGS73.

Table 2
Biology and General Science
Conversion Table

Number Right	Scaled Score	Number Right	Scaled Score	Number Right	Scaled Score	Number Right	Scaled Score
0	320	50	480	100	640	150	800
1	320	51	480	101	640	151	800
2	320	52	480	102	640	152	800
3	330	53	490	103	650	153	810
4	330	54	490	104	650	154	810
5	330	55	490	105	650	155	810
6	330	56	500	106	660	156	820
7	340	57	500	107	660	157	820
8	340	58	500	108	660	158	820
9	340	59	500	109	670	159	830
10	350	60	510	110	670	160	830
11	350	61	510	111	670		
12	350	62	510	112	670		
13	360	63	520	113	680		
14	360	64	520	114	680		
15	360	65	520	115	680		
16	370	66	530	116	690		
17	370	67	530	117	690		
18	370	68	530	118	690		
19	380	69	540	119	700		
20	380	70	540	120	700		
21	380	71	540	121	700		
22	390	72	550	122	710		
23	390	73	550	123	710		
24	390	74	550	124	710		
25	400	75	560	125	720		
26	400	76	560	126	720		
27	400	77	560	127	720		
28	410	78	570	128	730		
29	410	79	570	129	730		
30	410	80	570	130	730		
31	420	81	580	131	740		
32	420	82	580	132	740		
33	420	83	580	133	740		
34	420	84	590	134	750		
35	430	85	590	135	750		
36	430	86	590	136	750		
37	430	87	590	137	760		
38	440	88	600	138	760		
39	440	89	600	139	760		
40	440	90	600	140	760		
41	450	91	610	141	770		
42	450	92	610	142	770		
43	450	93	610	143	770		
44	460	94	620	144	780		
45	460	95	620	145	780		
46	460	96	620	146	780		
47	470	97	630	147	790		
48	470	98	630	148	790		
49	470	99	630	149	790		

Table 3

Content Categories of Biology and General Science

Frequency Distribution of Number of Questions in Each Category Correctly Answered by Percentages of Examinees

Percentage Ranges	History, Philosophy, and Methodolgy of Science I	Molecular and Cellular Biology II	Biology of Plants, Animals, Fungi, and Protists III	Evolution IV	Ecology V	Chemistry VI	Physics VII	Earth and Space Science VIII	Number of Questions Within each Range
↑ 95+			1						1
90-94	2	1		1	2		1		7
85-89	3	2	3	1	3	1		1	14
80-84	1	2	3	1	4	2	4	2	19
"Easy" 75-79	3	2	2	2		4	1	3	17
70-74	1	1	4	2	3	2	1	1	15
65-69	2	2	2	2	4	1	2		15
60-64		2	4	2	1	1		1	11
55-59	2	3	5	2		1	2	3	18
50-54		1	2	1	2	2		1	9
45-49		2	3		1			2	8
40-44	2	1	1	1		1	1	1	8
35-39		1	1	1					3
30-34		1	2		1	1	3	1	9
25-29		1		1			1		3
"Hard" 20-24		1							1
15-19		1		1					2
10-14									
↓ 5- 9									
Total Number of Questions	16	24	33	18	21	16	16	16	160
Mean Number Answered Correctly	11.6	14.2	20.8	11.0	15.1	10.6	9.8	10.1	103.1
Mean Number Answered Incorrectly *	4.4	9.6	11.8	6.9	5.9	5.2	6.0	5.8	55.6

* Questions that were not answered or were not attempted are not included as incorrect responses.

Estimating an Actual Score from Your Practice Score

When you actually take the Biology and General Science test, you will have questions to answer that will be similar to the questions in this book, but they will not be identical. Because of the differences in questions, the test you actually take may be slightly more or less difficult than the Biology and General Science test printed in this book. Therefore, you should not expect to get the same score that you have gotten on this test.

How to Interpret Your Score

The most meaningful interpretation of a test score occurs in the context of a particular use of the test. Thus, the interpretation of your score depends on the purpose for which you are taking the test. Scores may be interpreted in relation to some level of performance, such as a minimum score considered acceptable to satisfy established proficiency requirements. For example, if you are taking the Biology and General Science test because it is required for teacher certification in your state, the most useful information you could have would be the value of the qualifying score used by the teacher certifying agency in your state. Information about state qualifying scores or institutional requirements, which are determined by the state agencies and institutions, can be obtained directly from them.

Scores on the Biology and General Science test also may be interpreted in relation to the performance of appropriate reference groups of individuals who have taken the test. To assist you in interpreting your scaled score, percentile rank information based on data from the most recent three-year period of test administrations is provided in *Interpreting NTE Specialty Area Test Scores*, a leaflet that is sent with examinees' score reports. You can compare the score you obtained on the practice test in this book with the ones shown in the table to see if you scored above or below the average. The leaflet is updated yearly and may be obtained by writing to NTE Programs, Educational Testing Service, P.O. Box 6051, Princeton, NJ 08541-6051.

Interpretive Information for Test Questions and Categories

The table listing the correct answers (Table 1) also lists the percentage of examinees choosing each correct answer. Thus, you can compare your performance on each question with that of the March 1991 group. The percentages give an indication of whether the group as a whole found a question easy (say, 75 percent chose the correct answer) or difficult (say, 25 percent chose the correct answer).

Information based on the percentages from Table 1 is summarized in Table 3, which lists the number of questions in each content category that fall within percentage ranges, the total number of questions in each category, the average (arithmetic mean) number of questions answered correctly, and the average number of questions answered incorrectly. Thus, if you count the number of questions you answered correctly in each content category, you can compare how well you did with how the March 1991 group did in general on each category of the test.

You can also do a similar comparison based on the number of questions you answered incorrectly. When you examine Table 3, note that some questions or categories may appear to be more difficult than others. The apparent difficulty of the questions (or category) may be affected not only by the actual difficulty of the questions themselves, but also by the ability level and training of the group on which the statistics are based.

Explanations of the Answers to the Practice Test Questions

1. The release of an egg from the ovary is called

   (A) menstruation
   (B) ovulation
   (C) oogenesis
   (D) cleavage
   (E) fertilization

The five choices relate to the reproductive cycle and early development of the zygote. Choice B is the correct answer. Ovulation is the release of an egg from the ovary. Menstruation is the vaginal discharge of blood and tissue that results from the breakdown of the endometrium. Oogenesis is the production of female gametes. Fertilization takes place when two gametes fuse to produce a diploid zygote, and cleavage is the rapid division of cells that converts the zygote into a ball or a sheet of cells.

---

2. Sound waves travel fastest through a

   (A) vacuum
   (B) gas at high pressures
   (C) gas at low pressures
   (D) solid
   (E) liquid

Sound waves are defined as audible longitudinal disturbances in matter. Since a disturbance in matter is required, choice A, a vacuum, is incorrect. In a gas, sound travels at approximately 330 meters per second, in a liquid such as water, at 1,435 meters per second, and in a solid such as a metal, at 4,950 meters per second. Sound waves travel fastest through a solid, and choice D, therefore, is the correct answer.

---

3. In evolutionary history, which of the following classes of organisms is believed to have preceded the others?

   (A) Osteichthyes
   (B) Hydrozoa
   (C) Aves
   (D) Reptilia
   (E) Insecta

Organisms in the classes Insecta, Osteichthyes, Aves, and Reptilia have bilateral symmetry, a tubular mouth-to-anus gut, and a coelom. These characteristics are considered to be more advanced in terms of evolutionary history than the characteristics of organisms in the class Hydrozoa. Members of this class have radial symmetry and a simple body plan, that is, a sac with a central digestive cavity and a single opening serving as both mouth and anus. Thus, members of the class Hydrozoa are believed to have arisen before the organisms in the other four classes, and choice B is the best answer.

---

4. When white light hits a black object, the light is primarily

   (A) reflected
   (B) scattered
   (C) absorbed
   (D) refracted
   (E) diffracted

An object appears black primarily because it exhibits general absorption; that is, it absorbs all wavelengths more or less equally, and nearly all of the energy incident upon it is absorbed. When light is reflected, scattered, or refracted, some light would return from the object to the observer and thus the object would not appear to be black. Diffraction (diffracted) refers to the behavior of light on its passage through an aperture comparable in size to the wavelength of the light and would not be of relevance here. Thus, choice C is the correct answer.

---

5. The phases of the Moon are mainly due to the

   (A) distance of the Moon from the Earth
   (B) distance of the Moon from the Sun
   (C) rotation of the Earth on its axis
   (D) angle of the Moon's orbit to that of the Earth's orbit
   (E) position of the Earth and Moon relative to the Sun

The Moon does not emit its own light but reflects light received from the Sun. It is the position of the Earth and Moon relative to the Sun that determines the phase of the Moon. The half of the moon that faces the Sun is always lighted, and the phases that are seen from Earth are determined by how much of the lighted half is visible. The correct answer, therefore, is choice E, the position of the Earth and Moon relative to the Sun.

---

6. The attractive force that most strongly influences or governs the movement of astronomical bodies is called the

   (A) cohesive force
   (B) magnetic force
   (C) gravitational force
   (D) centripetal force
   (E) electrical force

Since the time of Newton, the motion of the astronomical bodies has been modeled successfully using gravitational force alone. The other forces, if they have any effect, are much smaller. Choice C, therefore, is the correct answer.

---

7. Which of the following is an advantage of using wood, rather than fossil fuels, to heat or supplement the heating of homes?

(A) Wood is a renewable natural resource.
(B) Wood provides more heating value, kilogram for kilogram, than coal, oil, or natural gas does.
(C) Wood is easier for people to obtain, transport, and store than fossil fuels are.
(D) Only wood contains compounds of hydrogen and carbon, which provide heating value.
(E) Wood does not produce air pollution.

Fossil fuels are nonrenewable sources of fuel. The fact that wood is a renewable natural resource is, therefore, an advantage of using wood rather than fossil fuels to heat or supplement the heating of homes. Wood has a low energy content per unit of weight and a high water content and does not provide more heating value. For most people, wood is more difficult to obtain, transport, and store than fossil fuels. With respect to choice D, since both fossil fuels and wood contain hydrogen and carbon, wood cannot be considered advantageous for this reason. The statement "Wood does not produce air pollution" is incorrect. When wood is burned, large quantities of carbon monoxide and other pollutants are given off. Thus choices B, C, D, and E are incorrect, and choice A is the correct answer.

8.

Time (hours)	Observed Counts per Minute
0	204
1	106
2	51
3	22
4	13

A radioactive source is monitored for one minute at hourly intervals by use of a Geiger counter, yielding the data above. All of the following statements are consistent with these data EXCEPT:

(A) Statistical fluctuations are present in the data.
(B) Doubling the observation times to two minutes will approximately double the number of counts recorded.
(C) The half-life of the source is approximately one hour.
(D) The count rate will increase in the coming hours.
(E) An hour before the first observation, approximately 400 counts would have been recorded during a one-minute observation.

Four of the five choices are consistent with the data given for the monitoring of a radioactive source over a period of time. Statistical fluctuations are present in the data. The counts per minute do not reduce by exactly one half each hour. If one were to double the time of observation, the counts would approximately double. The fact that the counts reduce by approximately one half each hour indicates that the half-life of the compound is approximately one hour. If a count was taken approximately one hour before the first observation, it appears from the data given that this count would be approximately 400. Choice D, an increasing count rate in the coming hours, would not be consistent with the data. Thus, choice D is the best answer.

---

9. Approximately how many years will it take for light from Sirius, the Dog Star, to reach Earth if Sirius is about 8×10^{13} kilometers distant from Earth?
(1 light-year = 9.46×10^{12} kilometers)

(A) 0.85 years
(B) 1.2 years
(C) 8.5 years
(D) 12 years
(E) 75 years

Astronomers use the unit light-year to measure distances in space. A light-year is defined as the distance light travels in one year and is equal to 9.46×10^{12} kilometers.

Since the number of years = $\frac{\text{distance in kilometers}}{\text{speed in kilometers per year}}$,

then the number of years = $\frac{8 \times 10^{13} \text{ kilometers}}{9.46 \times 10^{12} \text{ kilometers per year}}$,

which is 8.5 years.
Thus, choice C is the correct answer.

---

10. Prior to the appearance of photosynthetic organisms, the early atmosphere of the Earth is believed to have contained only trace amounts of

(A) N_2
(B) O_2
(C) NH_3
(D) CH_4
(E) H_2O

The early atmosphere of Earth was quite different from today's oxidizing atmosphere. It is thought to have been made up primarily of gases such as N_2, NH_3, CH_4, and H_2O that are still found in the gaseous emissions of volcanoes. O_2 is believed to have begun to accumulate when the process of photosynthesis evolved. This process converted the atmosphere to an oxidizing one. Choice B, therefore, is the best answer.

---

11. The addition of fertilizers to a lake will initially affect the oxygen cycle in the lake by

(A) promoting algal growth
(B) breaking down oxygen directly
(C) causing more oxygen from the air to be used at the lake's surface
(D) causing organisms to use less oxygen
(E) inhibiting oxygen production, causing an increase in carbon dioxide production

Choice A is the correct answer. Fertilizers contain nitrates, phosphates, and potassium in varying amounts. These nutrients allow for the greater production of crops. They can, however, cause excessive algal growth when allowed to pollute lakes. These algal blooms may result in increased oxygen production during the day, but they may also cause oxygen depletion at night due to respiration. As the algae die, decomposers break down the organic matter and deplete the oxygen.

---

12. All of the following represent examples of chemical changes EXCEPT

(A) maple leaves photosynthesizing
(B) wax melting
(C) wood burning
(D) silver tarnishing
(E) iron rusting

A chemical change is defined as a change of substances into other substances through the reorganization of atoms. Choices A, C, D, and E are all chemical changes. When maple leaves photosynthesize, carbon dioxide and water are converted to sugar through a series of chemical changes. When wood burns, it is converted to ash, and carbon dioxide and water are emitted. Silver tarnishes when silver sulfide forms on its surface, and iron rusts when oxygen and water are present and cause iron oxide to form. When wax melts, it is changing from a solid state to a liquid state. This is a physical change because its chemical composition is not altered. Thus, choice B is the correct answer.

---

13. Which of the following would be most useful for separating blood cells from plasma in order to study each component separately?

(A) Electron microscope
(B) Stereomicroscope
(C) Microdissection instruments
(D) Filter paper
(E) Centrifuge

An electron microscope and a stereomicroscope are both used to magnify objects. Microdissection instruments are used to dissect at the cellular level, and filter paper is not fine enough to separate blood cells effectively from plasma. When blood is centrifuged, the red blood cells will settle to the bottom of the centrifuge tubes and thus separate out. Therefore, choice E, the centrifuge, would be most useful for separating blood cells from plasma in order to study each component separately.

---

14. Translation directly involves all of the following EXCEPT

(A) rRNA
(B) tRNA
(C) mRNA
(D) amino acids
(E) DNA

During translation, the sequence of bases in mRNA is used to order the amino acids to form a polypeptide. The mRNA acts as a template for the synthesis of polypeptide chains. As the ribosomes (which are composed of rRNA and protein) move along the mRNA, they read the codons starting at the 5' end. Amino acids to be incorporated into the polypeptide chains are picked up by molecules of tRNA specific for each of the 20 amino acids. The tRNA's bring their amino acids to the ribosomes as they move along the mRNA. Each tRNA attaches to the mRNA at the point where a triplet of mRNA bases (a codon) is complementary to the exposed triplet (anticodon) on the tRNA. This ordering of the tRNA's along the mRNA automatically orders the amino acids. While DNA is the blueprint for the production of proteins, it is not directly involved in the mechanism of translation. Thus, choice E is the correct answer.

---

15. The Galapagos finches studied by Darwin are a good example of

(A) inheritance of acquired characteristics
(B) nonselection
(C) adaptive radiation
(D) speciation by polyploidy
(E) convergent evolution

Adaptive radiation is defined as the evolution of several different species from one ancestor, each species filling a distinct ecological niche. Darwin's finches, which inhabit the Galapagos Islands, are one of the best examples of this. Thirteen different species arose from a single ancestral type, believed to have been a ground-feeding finch. Thus, C is the correct choice. Inheritance of acquired characteristics is part of Jean-Baptiste Lamarck's concept of evolution, which with our current genetic knowledge is known to be incorrect. Nonselection clearly is incorrect as different species have been selected. Speciation by polyploidy has not taken place as the finches have no more than two sets of chromosomes per nucleus. Convergent evolution would assume that the finches are totally unrelated and have developed their similarities due to their environment.

---

16. When a homozygous red-flowering snapdragon is crossed with a homozygous white-flowering snapdragon, all the offspring produce pink flowers. This phenomenon is known as

(A) incomplete dominance
(B) sex-linkage
(C) pleiotropy
(D) recombination
(E) inversion

Sex-linkage refers to an inherited trait determined by an allele located on a sex chromosome, thus showing a different pattern of inheritance in males and females. Pleiotropy is the capacity of a gene to affect several different phenotypic characteristics. The only characteristic in this question is flower color. Recombination is the formation of new gene combinations, and inversion takes place when a segment of chromosome is reversed. Incomplete dominance can be seen when offspring show the effects of both alleles at a particular locus. The allele for red flowers and the allele for white flowers are both exerting equal effects, resulting in offspring with pink flowers. The correct answer, therefore, is choice A.

---

17. In a rocky region with no soil, the first stage of organisms in an ecological succession is usually

(A) mosses and ferns
(B) oaks and pines
(C) lichens and mosses
(D) oaks and beeches
(E) ferns and shrubs

Since no soil is present in the rocky region, organisms with extensive root systems cannot be supported. At least one organism in each pair in choices A, B, D, and E would be unlikely to be the first organism to invade the region. Lichens and mosses are both considered to be early colonists. Lichens, formed from cyanobacteria (blue-green algae) and fungi living together in a symbiotic relationship, are known as the first colonists. They can survive on rock and are found throughout the world from the Arctic to the Antarctic. Their activities begin the process of soil formation so that organisms such as mosses can anchor themselves using rhizoids. The correct answer, therefore, is choice C.

---

18. A recently captured young elephant trumpets loudly and sways from side to side when a person enters its stockade. This behavior continues but diminishes in intensity until it finally disappears. This change in behavior can be classified as

(A) habituation
(B) trial and error (operant conditioning)
(C) imprinting
(D) classical conditioning
(E) imitative learning

Choices A through E are all patterns of behavior associated with learning. After a period of time, the elephant is clearly not reacting to the person who enters its stockade each day. Operant conditioning occurs, for example, when an animal learns to run a maze and is rewarded by food. The elephant is not being rewarded for learning a task. Imprinting is learning that takes place during a specific critical period in an animal's life, such as a newly hatched duckling following the first moving object it sees, and that does not apply in this question. Classical conditioning occurs, for example, when a dog salivates at the ringing of a bell because it has learned to associate the bell with food. Imitative learning is learning by copying, such as the macaques (rhesus monkeys) in Japan learning from one individual how to wash sand off food. Habituation is considered to be one of the simplest forms of learning. The elephant becomes accustomed to a person routinely entering its stockade and learns over time to ignore the individual. Thus, choice A is the correct answer.

---

19. The superficial resemblance between a porpoise and a fish is due to

(A) adaptive radiation
(B) convergent evolution
(C) nonrandom mating
(D) gene migration
(E) genetic drift

Porpoises and fish are both in the phylum Chordata and subphylum Vertebrata, but they are in different classes. Their superficial likeness stems from the independent evolution of similarities due to living in similar environments. Thus, convergent evolution, choice B, is the correct choice. Adaptive radiation, the evolution of several different species from one ancestor, such as Darwin's finches, does not explain why porpoises and fish seem superficially to be similar. Nonrandom mating implies that in some way mating has been controlled to produce the similarities. This is clearly incorrect. Gene migration implies that genes from the porpoise moved to the fish, or vice versa, and caused the two organisms to look somewhat similar. Genetic drift, a change in allele frequencies occurring by chance, suggests that porpoises and fish should in fact look very dissimilar, and thus it is also incorrect.

---

20. All of the following are primary functions of the mammalian skeleton EXCEPT

(A) movement
(B) storage of fat and inorganic salts
(C) support and shape to the body
(D) heat regulation
(E) protection

The mammalian skeleton has four primary functions. Movement takes place because muscles are attached to the skeleton and thus move it. Some storage of fat and inorganic salts takes place in the bone. The skeleton gives support and shape to the body, and vital organs, such as the heart and the liver, are protected by the skeleton. The skeleton does not regulate heat. This is done by the hypothalamus, which is situated in the brain. Choice D, therefore, is the correct answer.

---

21. As a stream flows downhill, it carries a load of sediments. If the speed of the water decreases, which of the following occurs?

(A) The amount of erosion by the stream increases.
(B) The kinetic energy of the stream increases.
(C) The potential energy of the stream increases.
(D) Deposition of sediment by the stream increases.
(E) The particle size of sediment carried by the stream increases.

The water, as its speed decreases, will lose some of its kinetic energy (energy of motion), and its potential energy (stored energy) at points further downstream will decrease. Choices B and C are therefore incorrect because they state the opposite of what in fact happens. Because the total energy decreases, less energy is available to move sediment. When this occurs, the amount of erosion decreases, the size of sediment the stream is able to carry decreases, and the deposition of sediment by the stream increases. Thus, choices A and E are incorrect, and D is correct.

---

22. During a classroom discussion on the evolution of the giraffe's neck, a student remarked, "The giraffe stretched its neck while reaching for higher leaves, and its offspring had longer necks as a result." To correct the student's misconception, the teacher should emphasize which of the following?

(A) Overproduction of offspring leads to a struggle for survival.
(B) Only favorable adaptations have survival value.
(C) Spontaneous mutations can cause the sudden appearance of new traits.
(D) Disuse of an organ will often lead to its disappearance from a species.
(E) Characteristics acquired during one's lifetime are not passed on through the genes.

The student's misconception concerning the giraffe's neck is an outcome of Lamarck's theory of use and disuse, which states that characteristics acquired during a lifetime can be passed on to the next generation. Thus, according to Lamarck, the giraffe stretching its neck would result in a longer neck and this longer neck would be passed on to the next generation. In light of genetic studies, Lamarck's theory is now known to be false. Only heritable traits coded for by the genes on the chromosomes of gametes can be selected for and passed on to the next generation. The teacher would have to emphasize that traits acquired during one's lifetime are not passed on through the genes. The correct answer is choice E.

---

23. Houses have been built in a desert area. Trees, lawns, and flowers have been established with the aid of artificially supplied water. Which of the following would most likely occur if water were no longer supplied to the area?

(A) The area will go through secondary succession and the result likely will be a chaparral.
(B) The area will go through secondary succession and the result likely will be a deciduous forest.
(C) The land will return to desert.
(D) A grassland will form since the lawn grasses will remain after the trees die from a lack of water.
(E) Nothing will happen because houses are present.

Desert areas have very little rainfall, and the number of organisms that exist in these areas is accordingly small. Although vegetation artificially established might continue for a short while after the stoppage of water, eventually it will die because it lacks the adaptations of the indigenous organisms that live in low-moisture areas. Thus, the area will revert to its former state and the land will return to desert. The best answer, therefore, is choice C.

---

24. A substance that contains atoms with unstable nuclei is said to be

(A) ionic
(B) covalent
(C) radioactive
(D) amphoteric
(E) electronegative

Ionic and covalent both describe types of chemical bonds. An amphoteric substance is one that can act as an acid or as a base, and electronegative means carrying a negative charge of electricity. None of the above choices is the definition of a substance that contains unstable nuclei. A radioactive substance is one whose nuclei spontaneously disintegrate emitting particles and/or rays. Thus, choice C is the correct answer.

---

25. Scientists have proposed in recent years that the burning of fossil fuels increases the amount of CO_2 in the atmosphere. It has been predicted that if the level of CO_2 continues to rise, then the

(A) Earth's surface will have increased exposure to ultraviolet rays
(B) average annual surface temperature will increase enough to cause major ecological disruptions
(C) excess CO_2 concentration when inhaled will cause an adverse effect on animals
(D) Earth will no longer be able to absorb enough solar radiation to warm its surface
(E) ozone layer will not be able to absorb all of the extra CO_2

When fossil fuels are burned, carbon dioxide is given off. The average global levels of CO_2 in the atmosphere have increased about 26 percent from 1850-1986. When energy from the Sun reaches Earth, it is radiated back into space as infrared radiation or heat. The gases in the atmosphere reradiate some of the escaping heat back to Earth. As the concentration of gases such as CO_2 increases, more heat will be reradiated back to Earth and the average atmospheric temperature will increase. It has been predicted that this increase will cause major ecological disruptions. The correct answer is choice B.

---

26. In a large population of organisms that is in Hardy-Weinberg equilibrium, gene *A* is found in two allelic forms, *A* and *a*. If the frequency of the dominant allele *A* in the population is 0.7, approximately how many individuals out of a hundred would be expected to be homozygous recessive?

(A) 6
(B) 9
(C) 14
(D) 30
(E) 49

If there are only two alleles in a population and the frequency of one is known, the frequency of the other can be calculated. Knowing that $p + q = 1$, where *p* is the frequency of allele *A* and *q* is the frequency of allele *B*, then *p* is 0.7 and *q* is 0.3. The sum of the frequencies of all possible genotypes in a population must add up to 1.

p^2	+	$2pq$	+	$q^2 = 1$
frequency of *AA*		frequency of *Aa* or *aA*		frequency of *aa*

In the population given, *p* is 0.7 and p^2 is 0.49, which is the frequency of homozygous dominant individuals. The frequency of homozygous recessive individuals is q^2. As *q* is 0.3, q^2 must be 0.09. Therefore, 9 individuals out of a hundred must be homozygous recessive, and choice B is the correct answer.

---

27. Under the same environmental conditions, a rock broken into small pieces weathers faster than the same rock remaining in one large piece. This phenomenon is best accounted for by which of the following?

(A) The hardness of the rock is greater than the hardness of the pieces.
(B) The pieces have a greater total surface area than the large rock has.
(C) The heat generated from breaking the rock causes a chemical change in the pieces.
(D) Rocks in water weather faster than rocks on land do.
(E) All of the pieces together weigh less than the large rock does.

With respect to choice A, the hardness of the rock is the same whether it is in large or small pieces, and therefore hardness cannot account for the faster weathering. Choice C is incorrect. The amount of chemical change caused by the heat concomitant with the breaking of the rock would be negligible. Choice D states that rocks in water weather faster than rocks on land do, and yet the question clearly states "under the same environmental conditions." Choice E is not only an incorrect statement but is also irrelevant to the question being asked. Choice B is the correct answer. The larger surface area of the pieces means that there are more exposed areas for the rain and wind to attack, and therefore the rock that is split into smaller pieces will weather faster than the same rock remaining in one piece.

---

28. DNA molecules of different sizes can be separated in the laboratory by use of

(A) spectrophotometry
(B) immunofluorescence
(C) electrophoresis
(D) electron microscopy
(E) X-ray crystallography

The question asks how DNA molecules can be separated. Spectrophotometry is a technique used to measure the proportions of light of different wavelengths that are absorbed and transmitted by a solution. This technique, therefore, will not separate DNA molecules. Immunofluorescence uses fluorescence microscopy to locate antibodies or antigens in or on cells and is not used to separate different sized molecules of DNA. Electron microscopy is a technique used to view very small objects, not separate them. X-ray crystallography is used to determine the structure of compounds; it does not separate them. Thus, electrophoresis, choice C, is the best answer. The DNA is applied to a gel and subjected to an electric current. The DNA molecules will then separate according to size.

---

29. Three organisms are described below.

Organism 1 is bilaterally symmetrical, coelomate, shows protostome-type development, possesses a chitinous exoskeleton, and is parthenogenetic.
Organism 2 is bilaterally symmetrical in the larval form, coelomate, shows deuterostome-type development, possesses an internal skeleton made of calcium-containing plates and a water-vascular system, and is dioecious.
Organism 3 is bilaterally symmetrical, coelomate, shows protostome-type development, possesses a hydrostatic skeleton, and is monoecious (hermaphroditic).

Which of the following statements concerning phylogenetic relationships among the organisms is most likely correct?

(A) Organisms 1 and 2 are more closely related to each other than either is to organism 3.
(B) Organisms 1 and 3 are more closely related to each other than either is to organism 2.
(C) Organisms 2 and 3 are more closely related to each other than either is to organism 1.
(D) Organisms 1, 2, and 3 are all equally closely related.
(E) There is not enough information to distinguish phylogenetically between organisms 1, 2, and 3.

The three organisms possess coeloms. The phyla of coelomate animals can be divided into two groups, based on early embryonic development. The egg after fertilization will either divide spirally, with the mouth developing at or near the blastopore (protostome), or radially, with the anus forming at or near the blastopore (deuterostome). Thus, organisms 1 and 3 are more closely related to each other than either of them is to organism 2, because they both show protostome-type development. Thus, choice B is the correct answer.

30.

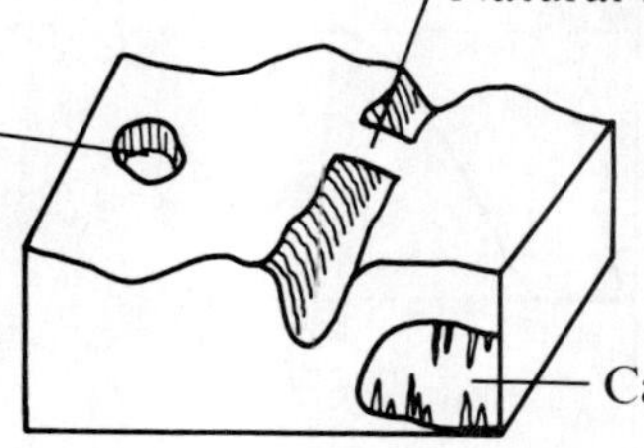

Which of the following correctly identifies the one factor that would result in the formation of the three erosional features indicated in the diagram above?

(A) Groundwater
(B) Volcano
(C) Wind
(D) Runoff
(E) Earthquake

Carbonate rocks such as limestone frequently have fissures running through them. Groundwater contains dissolved carbon dioxide, which produces an acidic solution. As the groundwater seeps through the cracks and fissures, it dissolves the limestone and carries it away, causing caves to be formed. After tens of thousands of years, surface areas called sinkholes form from collapsed cave passages. Natural bridges can also be formed when a surface river disappears into a fissure and reemerges. Thus, the three erosional features shown in the diagram are caused by groundwater, and choice A is the correct answer.

31.

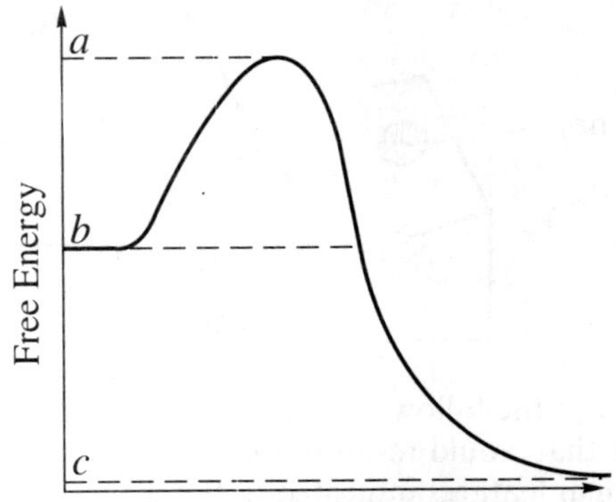

From the graph above, all of the following information concerning an enzyme reaction is available EXCEPT the

(A) energy absorbed by the enzyme
(B) energy content of the enzyme-substrate complex
(C) potential energy of the substrate molecule
(D) activation energy for the reaction
(E) energy released during the reaction

The graph depicts the progress of a reaction in the presence of a catalyst such as an enzyme. The substrate has an initial chemical potential energy (*b*) that can be determined from the *y*-axis. The energy content of the enzyme-substrate complex (*a*) can also be determined. The activation energy is calculated from *a* - *b* and the energy released during the reaction is *b* - *c*. The energy absorbed by the enzyme is impossible to deduce from the graph. Thus, choice A is the correct answer.

---

32. In the United States, most of the region east of the Mississippi River is occupied by which of the following biomes?

(A) Tundra
(B) Taiga
(C) Temperate deciduous forest
(D) Tropical rain forest
(E) Grassland

The region east of the Mississippi River in the United States has a warm, mild growing season with moderate rainfall. This is followed by a cold spell during which very little growth occurs. This climate is optimal for a temperate deciduous forest biome because the deciduous trees lose their leaves during a cold spell. Choice C, therefore, is the correct answer. Tundra, found at the northernmost limits of plant growth and at high altitudes, contains low shrubs or matlike vegetation. Taiga occupies most of Canada and the Soviet Union. Conifers grow there. Tropical rain forests are found along the equator. Grassland is found in central North America.

33.

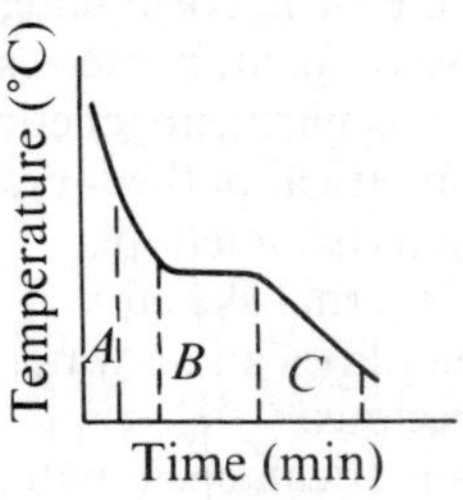

Substance *X*

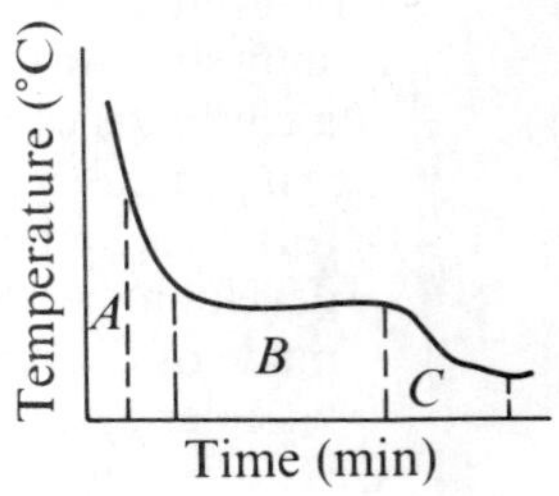

Substance *Y*

When the data obtained from either the cooling or melting of a substance are graphed, the presence of a plateau indicates an interval during which the liquid form of the substance is freezing or the solid form is melting without a change in temperature. The graphs above represent data from an experiment comparing the cooling of liquids *X* and *Y*. Based on the information above, all of the following statements are correct EXCEPT:

(A) During interval *A*, substances *X* and *Y* exist in the liquid state.
(B) During interval *C*, substances *X* and *Y* exist in the solid state.
(C) During interval *B*, substance *X* changes from a liquid to a solid without continued cooling.
(D) During interval *B*, substance *Y* changes from a liquid to a solid without continued cooling.
(E) Samples of *X* and *Y* have equal masses.

The cooling curves of two substances, *X* and *Y*, are shown in this question. Information regarding the states in which these substances exist at different temperatures can be found from the graphs. As the plateau represents the transition point between liquid and solid, choice A, which states that during interval *A*, substances *X* and *Y* exist in the liquid state, must be correct. Similarly, choice B must be correct, because the plateau has been surpassed and the temperature is decreasing; thus, solids must exist. The question states that the plateau indicates an interval during which the liquid form of the substance is freezing without a change in temperature; thus, choices C and D must be correct. There is no data given about masses; hence, the statement in choice E is incorrect. This exception is, therefore, the correct answer.

---

34. Which of the following represents the correct electron-dot formula for methane, CH_4?

(A) 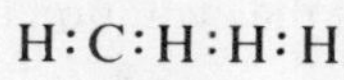

(B) 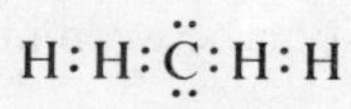

(C) 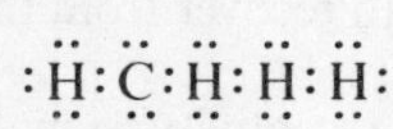

(D) 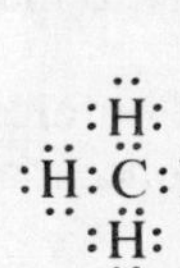

(E) 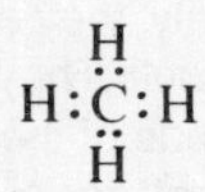

Electron-dot structures depict valence electrons as dots. The inner electrons and the atomic nucleus are represented by the symbol for the element being considered. Carbon has four valence electrons and hydrogen has one valence electron. Carbon and hydrogen can form covalent bonds with each other, allowing each element to achieve a stable electron configuration. Choices A and E have the correct number of electrons (8); however, carbon has four electrons to share and will do so with each of four hydrogen atoms, thereby forming four covalent bonds. Thus, choice E is the correct answer.

---

35. Which of the following major scientific pronouncements was proposed first?

(A) Koch's postulates concerning diseases and their causative agents
(B) Oparin's hypothesis about the origin of life
(C) Darwin's theory of evolution
(D) Watson and Crick's model of DNA
(E) Beadle and Tatum's one gene-one enzyme hypothesis

Darwin's theory of evolution was put forward in 1859; Koch's postulates concerning diseases and their causative agents was proposed in 1882; Oparin's hypothesis about the origin of life was proposed in the 1920's; Beadle and Tatum's one gene-one enzyme hypothesis was put forward in 1941; and the Watson-Crick model of DNA was proposed in 1953. The major scientific pronouncement that preceded the others, therefore, is choice C.

36.

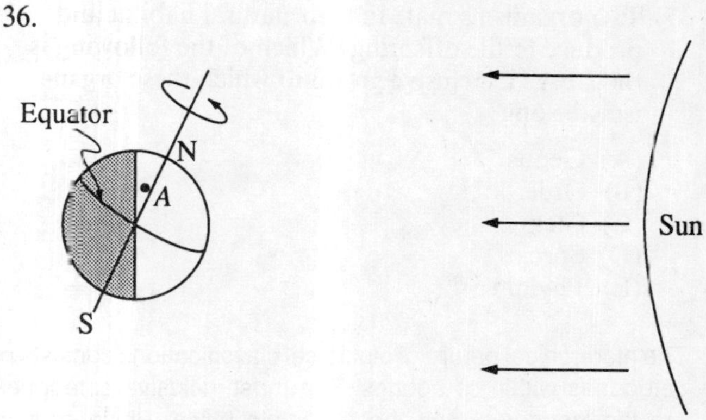

The diagram above shows one position of the Earth in relation to the Sun. When the Earth is in this position, what is the season of the year and the time of day at point *A* on the Earth's surface?

(A) Summer and morning
(B) Summer and noontime
(C) Summer and afternoon
(D) Winter and morning
(E) Winter and afternoon

As can be seen from the diagram, the Earth's axis is not perpendicular to the plane of its orbit around the Sun. It is tilted at a 23.5-degree angle from the perpendicular. It is this tilt that causes the seasons. As the Earth travels around the Sun, the axis always points in the same direction; therefore, on one specific day of the year the North Pole is tilted 23.5 degrees toward the Sun, and six months later it is tilted 23.5 degrees away from the Sun. In the summer the Northern Hemisphere is leaning toward the Sun and in the winter it is leaning away from the Sun. Thus, the season of the year shown in the diagram is summer. As the Earth travels around the Sun, it rotates on its own axis in a counterclockwise direction as viewed from the North Pole (one full rotation every 24 hours). Point A, therefore, has just moved into the light, so it must be morning. Thus, the correct answer to the question is choice A, summer and morning.

---

37. Two organisms mate in their natural habitat and produce fertile offspring. Which of the following is the LEAST inclusive group to which these organisms belong?

(A) Genus
(B) Order
(C) Class
(D) Species
(E) Phylum

The hierarchical nature of biological classifications consists of categories within categories. The most inclusive categories are the kingdoms, and these contain phyla. Phyla contain classes, classes contain families, families contain genera, and genera contain species. The question asks which of the choices is the least inclusive taxonomic group to which two organisms that mate in their natural habitat and produce fertile offspring can belong. This is the definition of a species and since species is the least inclusive taxonomic group, choice D is the correct answer.

---

38. The reason that a hypothesis is easier to disprove than to prove is that

(A) formulating a hypothesis is the first step in the application of scientific thinking to a problem; disproving as a second step is easier than the first step
(B) a hypothesis can become a law only if it is proved to be true in at least a dozen different research centers; it can be disproved at a single center
(C) measurements with precision can disprove a hypothesis; measurements must have both accuracy and precision to prove a hypothesis
(D) a hypothesis can be disproved by one person but can only be developed by many scientists working together
(E) one experiment may disprove a hypothesis but many measurements cannot show that a hypothesis is always true

A hypothesis is defined as a temporary working explanation of a problem. If it is not validated, it has to be discarded. A hypothesis can be supported by the results of many experiments; however, one experiment with reproducible results may be enough to disprove it. Thus, choice E is the best answer to the statement that a hypothesis is easier to disprove than prove. One experiment may disprove a hypothesis, but many are needed to support it.

---

39. At about three years of age, male lions leave the pride in which they were born. After a period of nomadic existence, each eventually takes over a pride of its own. What is the evolutionary advantage of this behavior pattern?

(A) It increases competition between males of the same pride.
(B) It permits females to recover from their reproductive event.
(C) It enhances female receptivity to male mating behavior.
(D) It increases the likelihood of outbreeding.
(E) It encourages pair-bonding once males and females reunite.

As young male lions approach maturity, they are driven from their pride by the established male adult. They tend to live in small groups until they take over a pride of their own, usually defeating an adult male. This type of social system helps ensure outbreeding and encourages new genetic combinations that may be advantageous in evolutionary terms. Thus, choice D is the correct answer.

---

40. If a ball is dropped from a tall tower, and at the same time a bullet is fired horizontally from the same location on the tower, the ball and bullet hit the ground at the same time.

For the statement above to be true, conditions that must be met include which of the following?

I. The ground must be level.
II. Air resistance must be absent.
III. The ball and the bullet must have the same mass.

(A) I only
(B) III only
(C) I and II only
(D) II and III only
(E) I, II, and III

The only force acting on either the ball or the bullet is the force due to gravity. Thus, both objects will experience the same vertical (downward) acceleration, regardless of the initial horizontal velocity of the bullet and regardless of the difference in their masses. It follows that the objects will hit the ground at the same time provided that the ground is level (since they will be hitting the ground at different horizontal locations) and provided that air resistance can be neglected. Conditions I and II only must be met and choice C, therefore, is the correct answer.

---

41. Which of the following factors help(s) to account for the fact that, in the United States, lands to the east of a mountain range receive less rain than lands to the west of the mountains receive?

I. Winds often blow from west to east.
II. An air mass cools as it goes over a mountain range.
III. A sample of cold air can contain less water vapor than a sample of warm air can.

(A) II only
(B) III only
(C) I and II only
(D) II and III only
(E) I, II, and III

The Earth's surface is covered by belts of air currents, which largely determine the movement of weather. In the United States, the westerlies are prevalent from the latitudes 30° N to 60° N; thus, statement I is correct and important in the determination of where rain falls. Statement II is also correct. As the air mass cools, it releases most of its water, resulting in rain on the windward (western) side of the mountain. As air masses travel from west to east, therefore, the land to the east of the mountain range will receive less rain than those lands to the west. Statement III is also true and relevant. As the cooled air passes over the mountain range, it will lose the water vapor as raindrops. The factors, therefore, that help account for lands to the east of a mountain range receiving less rain than those to the west of a mountain range are statements I, II, and III. Choice E is the correct answer.

---

42. The gene pool of a population of *Rana catesbiana* living in a pond has remained constant and stable for many generations. Which of the following is the most likely reason for this stability?

(A) The pond ecosystem experienced rapid succession and then reached equilibrium.
(B) No mutations occurred in a large migrating population.
(C) No migration occurred in a large population that was mating randomly.
(D) Random mating occurred in a small population that had a high frequency of gene mutations.
(E) Nonrandom mating occurred in a small population that had a high frequency of gene mutations.

For a gene pool to remain constant and stable for many generations, five conditions must be met. These conditions, put forward by both G. H. Hardy and W. Weinberg, are as follows: No net mutational change should take place; no immigration or emigration should take place; the population should be large enough for the laws of probability to apply; mating should be random; and there should be no selection for the genes in question. Choices A, B, D, and E have aspects of the five conditions, but each is inaccurate. For example, in choice B, "No mutations occurred" is a correct condition; however, "in a large migrating population" is incorrect because the condition to be met states that no immigration or emigration should take place. Choice C incorporates three of the conditions accurately and, therefore, is the correct answer.

---

43. Toxic materials are often found in high concentrations in the tissues of organisms at the top of food pyramids because these materials are first

(A) degraded in plants
(B) metabolized by plants
(C) excreted as wastes by animals
(D) accumulated in the cells of autotrophs and herbivores
(E) absorbed from the abiotic environment by carnivores

When organisms are taking in water and nutrients, they may accumulate toxic compounds in their tissues. The organisms cannot break these compounds down or excrete them. When these organisms are eaten, biological magnification takes place because the biomass at each trophic level is being produced from a much larger biomass at a lower trophic level. Thus, the chemicals not only become more concentrated, but are also found in numerous organisms due to the interconnection of the food chains. When autotrophs and herbivores accumulate the toxic nonfood materials in their cells and are in turn eaten by omnivores and carnivores, the toxic chemicals will then accumulate in the latter. Choice D, therefore, is the correct answer.

44.

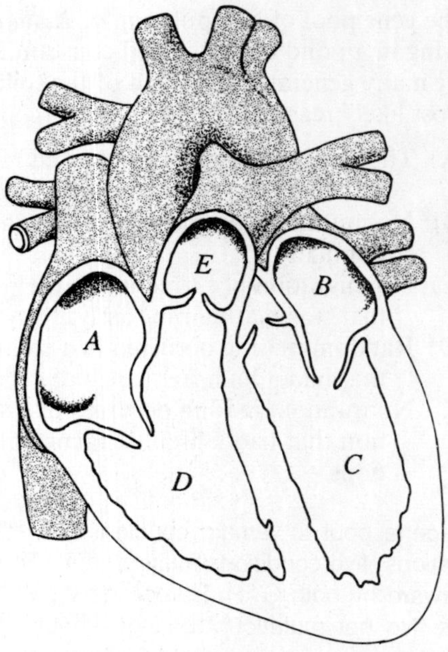

In the figure shown above, blood leaving the human heart and passing to the coronary circulatory system is expected to empty from which region?

(A) *A*
(B) *B*
(C) *C*
(D) *D*
(E) *E*

In the figure of the heart, *A* is the right atrium and *B* is the left atrium; *C* is the left ventricle and *D* is the right ventricle; and *E* is the pulmonary artery. The blood, traveling from the heart to the coronary system, will be oxygenated and will thus have come from the lungs via the pulmonary vein and entered the left atrium. It will travel from the left atrium into the left ventricle and go from there to the coronary circulatory system. Thus, choice C is the correct answer.

---

45. In the life cycle of the dog tapeworm, the flea, which can be assumed to be the only obligate intermediate host, gets the eggs from a dog's feces. When a dog eats the flea, the tapeworm larvae mature and mate in the intestine of the dog. Which of the following statements is correct about the eradication of the dog tapeworm?

(A) It could be accomplished by the elimination of fleas.
(B) It would require both the elimination of fleas and the ingestion of antitapeworm medication by all dogs.
(C) It would require both the ingestion of antitapeworm medication by all dogs and the sterilization of dog feces.
(D) It would require the elimination of fleas, the ingestion of antitapeworm medication by all dogs, and the sterilization of dog feces.
(E) It is impossible to achieve because of the complicated life cycle of the dog tapeworm.

The question states that the flea is assumed to be the only obligate intermediate host of the dog tapeworm. The term obligate intermediate host means that the tapeworm will not complete its life cycle unless it spends part of it within a flea. By eradicating the flea only, the tapeworm will not complete its life cycle. Choice A is the correct answer.

---

46. Monarch butterflies are distasteful to blue jays. Viceroy butterflies look very much like monarchs and consequently are not eaten by blue jays. This example best illustrates the phenomenon of

(A) mimicry
(B) altruism
(C) convergent evolution
(D) genetic drift
(E) the Hardy-Weinberg principle

The caterpillars of monarch butterflies feed on milkweed and store a toxic glycoside that the milkweed contains. The glycoside is then found in the monarch butterfly, making it unpleasant to eat. Monarchs also have very conspicuous colors that make them recognizable to predators. These predators, having once tasted the monarchs, will leave them alone because of their unpleasant taste. The viceroy butterflies are very similar to the monarchs in appearance. Predators have learned to avoid butterflies resembling monarchs. This is a classical example of mimicry. Thus, choice A is the correct answer.

---

47. Which of the following methods of insect pest control is potentially the most damaging to the environment?

(A) Baiting traps with the insect pest's pheromones
(B) Applying organic broad-spectrum pesticides to the areas being protected
(C) Genetically engineering crops that are resistant or immune to the insect
(D) Using a species-specific virus to attack the pest
(E) Using radiation to sterilize laboratory-raised male insects and releasing them to compete with wild males

The pest control methods in choices A, D, and E are all specific for the insect and will cause little damage to the environment. Making the crops themselves resistant to a particular insect, choice C, is less likely to cause environmental damage than applying organic broad-spectrum pesticides to the areas being protected. Thus, choice B is the correct answer. The chemicals will not only kill the insect pest but will also kill many beneficial insects. The pesticides may also enter the food chain.

---

48. Two structures that are found in the amniotic egg and that persist in the development of placental mammals are the

(A) chorion and the placenta
(B) yolk sac and the shell
(C) allantois and the umbilical cord
(D) chorion and the amnion
(E) vitelline membrane and the amnion

The amniotic egg, which is very similar to a hen's egg, first appeared in the reptiles. Only choice D has two structures that are found in the amniotic egg and that persist during the development of placental mammals. The chorion is the outermost of the four extraembryonic membranes of developing reptiles and contributes to the formation of the mammalian placenta. The amnion is the innermost of the four extraembryonic membranes and surrounds the embryo forming a fluid-filled chamber. Both the reptile embryo and the mammalian embryo float in this chamber, the fluid providing support for the embryos' tissues. Thus, choice D is correct.

---

49. In living organisms, the breakdown of polymers into monomers most often occurs by

(A) hydrolysis
(B) condensation
(C) dehydration
(D) anabolism
(E) photolysis

Metabolism can be divided into anabolism, the formation of complex substances from simpler ones, and catabolism, the breakdown of complex substances into simpler ones. Anabolism, therefore, is the opposite of what the question is asking. Condensation and dehydration are two terms for the same reaction, namely, the formation of polymers from monomers by the removal of water; thus they are also types of anabolism. Photolysis and hydrolysis are both catabolic. Photolysis is the decomposition or dissociation of a molecule as a result of the absorption of light, and hydrolysis splits one molecule (the polymer) into two or more molecules (monomers) by the addition of H^+ and OH^- from water. Thus, the correct answer is choice A.

---

50. Which of the following best describes the mass and the weight of a woman on the Moon, compared to her mass and weight on the Earth?

	Mass on Moon Compared to Earth	Weight on Moon Compared to Earth
(A)	1/36	1/36
(B)	1/6	1/36
(C)	Less	Same
(D)	Same	More
(E)	Same	Less

Mass is defined as the amount of matter that an object contains, whereas weight is a measure of the force of attraction between a body such as the Earth and the object. Since the mass is constant, choices A, B, and C, are incorrect. Weight is proportional to the mass of the attracting object. Thus, the weight of a woman on the Moon would be less than the weight of the same woman on Earth. Choice E, therefore, is the correct answer.

---

51. Suppose that the arrow end of a magnetized needle is a south pole. In which of the following diagrams does the arrow indicate the correct magnetic field direction? (N and S stand for north and south magnetic poles, respectively.)

(A)

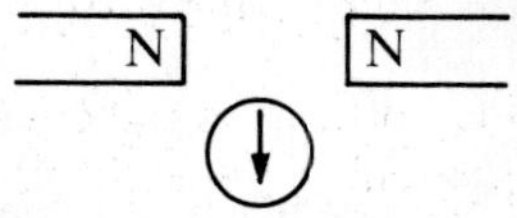

(B)

(C)

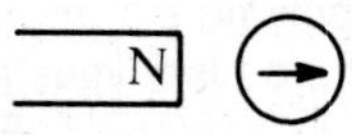

(D)

(E)

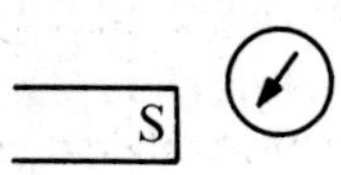

The magnetic force between two objects is due to their magnetic fields repelling or attracting each other. Magnets have polarity. The end of a suspended magnet that points north is the north-seeking pole, and the end that points south is the south-seeking pole. One of the properties of magnets is that like poles repel each other and unlike poles attract each other. In the diagrams, the head of the magnetized needle is the south pole or north-seeking pole, and it should be attracted by a north pole. The only choice that shows this attraction occurring is choice D, which is the correct answer.

---

52. Although adenosine triphosphate (ATP) is continually utilized during exercise, the temperature in the body does not change appreciably. This is an example of

(A) oxidation
(B) photophosphorylation
(C) homeostasis
(D) growth
(E) glycolysis

Adenosine triphosphate (ATP) is a molecule that provides energy for the cell as the ATP is hydrolyzed. Some of this energy is in the form of heat. In homeotherms (birds and mammals), body temperature is kept constant by a complex series of reactions that are controlled by the hypothalamus in the brain. For example, when the body gets too hot, blood vessels dilate and profuse sweating takes place. These responses serve to rid the body of the excess heat. This maintenance of a constant body temperature is not an example of either oxidation or glycolysis, which are both concerned with the breakdown of food substances. Photophosphorylation is a process that takes place during photosynthesis in which ATP is formed from ADP and has nothing to do with what the question is asking. Growth takes place continually, and maintenance of body temperature is obviously not an example of this process. Homeostasis is defined as the maintenance of a relatively stable internal environment, despite changing external conditions. The temperature in the body is not changing appreciably even with exercise; thus it is an example of homeostasis. Choice C is correct.

---

53. Characteristics of both mosses and liverworts include which of the following?

(A) A dominant sporophyte generation
(B) True roots, stems, and leaves
(C) Water-dependent fertilization
(D) A diploid (2n) chromosome number in the gametophyte generation
(E) A haploid (1n) chromosome number in the sporophyte generation

Mosses and liverworts are found in the division Bryophyta. Because they lack roots and other specialized tissue, choice B is incorrect. Mosses and liverworts both show alternation of generations; however, the gametophyte generation is the dominant generation, not the sporophyte generation; thus, choice A is incorrect. The gametophyte generation, which produces gametes, contains the haploid number of chromosomes; therefore, choice D is incorrect. Similarly, the sporophyte generation that forms after fertilization is diploid, making choice E a wrong answer. Both mosses and liverworts have motile sperm that require water in order to reach the eggs in the archegonia. Thus, choice C, water-dependent fertilization, is a characteristic that both mosses and liverworts have in common and is the correct answer.

---

54. Apical dominance, increases in the plasticity of cell walls, and the rapid elongation of cells are controlled by which of the following plant hormones?

(A) Auxins
(B) Gibberellins
(C) Cytokinins
(D) Ethylene
(E) Abscisic acid

Hormones are molecules that are secreted in one part of an organism that regulate functions in another part of the organism. Although all of the substances in the question are plant hormones, only auxins control apical dominance, increase the plasticity of cell walls by causing the cross-links between the cellulose molecules to break, and cause the rapid elongation of cells. Choice A, therefore, is the correct answer.

---

55. The essential amino acids must be obtained from food in the human diet because these amino acids

(A) form peptide bonds more easily than other amino acids do
(B) are found in special sequences of human proteins
(C) are more common in human proteins than in other proteins
(D) are not synthesized by the human body
(E) cannot be digested by humans

Amino acids are used by all organisms to make proteins. Humans are able to synthesize some amino acids but not others. Those not synthesized must be obtained by ingestion and are termed essential amino acids. Choice D is the correct answer.

---

56. Since color blindness is a sex-linked (X-linked) trait in humans, a man who is color-blind will pass on the allele for color blindness to

(A) half of his sons only
(B) half of his daughters only
(C) half of his sons and half of his daughters
(D) all of his sons
(E) all of his daughters

Color blindness is a sex-linked trait. A man who is color-blind will have the genotype X^cY, where c represents color blindness. One half of his gametes will contain the X chromosome, which carries the allele for color blindness, and one half will contain the Y chromosome. In humans the Y chromosome determines male sexual differentiation. Therefore, offspring receiving copies of the man's Y chromosome will be sons and will not receive the allele for color blindness. All of his daughters, however, will receive copies of his X chromosome, which carries the allele in question. Choice E, therefore, is the correct answer.

57. All of the following statements relating to gene mutations are valid EXCEPT:

(A) Gene mutations involve an alteration in the nitrogen-base sequence of DNA.
(B) Gene mutations are usually inherited as dominant alleles.
(C) Gene mutations may occur spontaneously, involving no apparent mutagenic agent.
(D) A change in only one nitrogen base may be sufficient to cause the production of a defective protein.
(E) Only those mutations occurring in germ cells can be passed along to future generations.

A mutation is defined as a change in the DNA sequence of an organism. Mutations may involve a change in the nitrogen-base sequence of DNA. They can seem to occur spontaneously, involving no known mutagenic agent. A change in one of the bases can lead to the production of a defective protein because the codon of the resulting mRNA can change the results of translation. Choice E is also true of gene mutations. If the DNA in gametes is altered and fertilization takes place, then the zygote will contain the changed DNA. Choice B, "Gene mutations are usually inherited as dominant alleles," is incorrect. Gene mutations may be recessive, sex-linked, or even "silent." Thus, choice B is the correct answer.

---

58. The goal of a teaching unit is for students to recognize that species are replaced by other species as their environment changes. Which of the following is an appropriate learning experience?

(A) Comparing plant types of a primary dune to those of a secondary dune
(B) Characterizing the stages of embryo development of a chick
(C) Observing phototropic movements by plants
(D) Recording changes in heart rate of *Daphnia* at different temperatures
(E) Comparing the brain structure and function among higher vertebrates

As the environment changes, species change. The species that are dominant at any given time are normally those species most adapted to particular environmental conditions. Because B, C, D, and E have nothing to do with species replacing other species, these would not be appropriate choices. With respect to choice A, comparing plant types of both primary and secondary sand dunes should enable students to see ecological succession in action and would be an appropriate learning experience for a unit on species replacing other species as their environment changes. Thus, choice A is correct.

---

59. All of the following statements relating to the fossil record are correct EXCEPT:

(A) The most ancient fossils are situated in the oldest layers of rock.
(B) Aquatic organisms evolved before land organisms.
(C) Many species have become extinct over time.
(D) Organisms have diversified over time.
(E) Dinosaur and human remains are usually found within the same rock layer.

Fossils are defined as the remains of an organism or as direct evidence of an organism's presence. They supply a record of organisms that were present on Earth, stretching back 3 billion years. It has been found through various dating techniques that the most ancient fossils are situated in the oldest layers of rock; that aquatic life forms preceded land organisms; that many species, for example, *Archaeopteryx*, are now extinct; and that organisms became more complex and diverse as time proceeded. Dinosaurs are believed to have become extinct about 65 million years ago, whereas the first humanlike primates are thought to have appeared around 5 million years ago. Thus, the remains of both would not be found within the same rock layers because this indicates the same relative age. Choice E is the correct answer.

60. In many animal societies, a social hierarchy, or peck order, controls the dominance behavior of its members. This peck order generally serves to accomplish all the following EXCEPT

(A) reduce aggression among the members of the society
(B) establish a means by which the members of the society obtain mates
(C) influence the way in which the members of the society distribute themselves within the living space
(D) establish a means by which food is distributed among the members of the society
(E) provide a means whereby the weak members of the society are protected and cared for

One type of dominance hierarchy found in birds and mammals is peck order. It is established when a group of animals such as hens are kept together over a period of time. A period of agonistic behavior takes place before the order is established. Once established, it does reduce aggression. It establishes a means by which the members of the society obtain mates: those high in the order will mate more often. It influences the way the members acquire space and food for themselves. The weak members, however, are not protected and cared for. They are literally lowest in the peck order. Thus, choice E is the correct answer.

61. An atmospheric relative humidity of 50 percent means that

(A) there is a 50 percent probability of rain
(B) half of the moisture in the air is condensing in the form of dew and droplets on objects
(C) the atmosphere contains 50 grams of water vapor per cubic meter
(D) the atmosphere is saturated with water vapor when the air temperature is 50° F
(E) the atmosphere contains half as much water vapor as it could contain at its present temperature

Relative humidity is defined as a measure of how close to saturation an air sample is at a specific temperature. It is expressed as a percentage. Thus, an atmospheric relative humidity of 50 percent means that the air sample is 50 percent saturated with water vapor. It is not an indication of the probability of rain and says nothing about condensing in the form of dew and droplets. Relative humidity is not measured in grams per cubic meter, and 50 percent relative humidity does not mean that the atmosphere is saturated with water vapor at 50°F. Choice E states that the atmosphere contains half as much water vapor (i.e., 50 percent) as it could contain at its present temperature and, according to the definition for relative humidity, is the correct answer.

62. Similarities in the bone structure of the foreleg of a horse, the wing of a bat, and the forelimb of a whale are examples of

(A) inheritance from a common ancestor
(B) inheritance of acquired characteristics
(C) evolution by use and disuse
(D) geographic distribution of species
(E) convergent evolution

Inheritance of acquired characteristics and evolution by use and disuse are both aspects of Lamarck's erroneous theory of evolution and are incorrect answers. Geographic distribution of characteristics is not relevant to the question, and convergent evolution refers to the independent development of similarities that result from adaptation to similar environments. The horse, bat, and whale do not live in similar environments. Thus, choice A, "inheritance from a common ancestor," is the correct answer. When the bone structure of the foreleg of a horse, the wing of a bat, and the forelimb of a whale are compared, they all show fundamental similarities of structure and organization.

63. In the Antarctic Ocean, fish have been found that can withstand freezing temperatures because of their ability to produce antifreeze compounds in their tissues. The accepted explanation for this phenomenon is that the

(A) cold temperatures caused ancestors of the fish to produce the compounds
(B) materials needed to form the compounds are available only in cold oceans
(C) fish that did not produce the compounds migrated to other areas
(D) isolation prevented other types of fish from entering the area
(E) fish that produced the compounds survived and reproduced

The fish that produce antifreeze compounds are adapted to their environment. At some point in their evolution a mutation, or a series of mutations, took place that enabled them to produce the compound that allowed them to survive and pass on this ability to their offspring, choice E. The cold temperatures would not have caused an ancestor of the fish to produce the compounds, but would have selected for those individuals able to produce them. Choice B is also incorrect, since the compounds are produced by the fish using materials present in food. Neither choice C nor choice D explains why some fish in the Antarctic Ocean produce the antifreeze compounds. All fish can enter or leave the ocean; whether they can survive or not depends on other factors. The fact that fish have been found that contain the antifreeze compound means that they can survive in severe conditions and also produce offspring. Thus, choice E is the correct answer.

---

64. Evidence from lakes in southeastern Canada and the northeastern United States indicates that lakes are becoming depleted of living organisms. Which of the following is considered the primary cause?

(A) Nuclear waste
(B) Lowered water levels
(C) Acid rain
(D) Garbage dumping
(E) Oil spills

Many lakes in southeastern Canada and the northeastern United States are becoming depleted of living organisms because the lakes now have a very low pH. This low pH renders the lakes intolerable for many forms of aquatic life. The lakes become acidic because sulfur oxides and nitrogen oxides emitted from coal- and oil-burning power stations in the central and midwest portions of the United States are being blown northeastward by the prevailing winds. The gases combine with water to form acid that falls as rain, hence the name acid rain. Thus, choice C is the correct answer.

---

65. The explosive growth and the success of rabbits in Australia after their introduction are due mainly to

(A) disruptive selection
(B) stabilizing selection
(C) directional selection
(D) reduced selection pressure
(E) isolation

Rabbits are not indigenous to Australia; they were introduced in 1859. Because they had no predators (except for humans), their growth in number was explosive. Disruptive selection is a process that favors phenotypic extremes, stabilizing selection eliminates extreme individuals, and directional selection results in an increase in the proportions of individuals with extreme phenotypic characteristics. None of these choices explains the success of the rabbit in Australia. Nor does isolation, which suggests that genetic isolation is causing the explosion, when in fact this refers to the lack of genetic exchange between species. Only one species was introduced. Reduced selection pressure, choice D, is the correct answer. There were very few pressures put on the rabbits; thus, most of the offspring survived and reproduced.

---

66. Which of the following types of behavior is manifested by a male redwing blackbird who attacks any blackbird that approaches the area where he is displaying?

(A) Dominance
(B) Latent learning
(C) Conditioned reflex
(D) Appeasement
(E) Territoriality

Breeding territories are established by many birds. Within these areas, they carry out courtship, build nests, and raise their young. A male redwing blackbird that attacks any other blackbird that approaches his area while he is displaying is manifesting a type of behavior known as territoriality. Choice E, therefore, is the correct answer.

67.

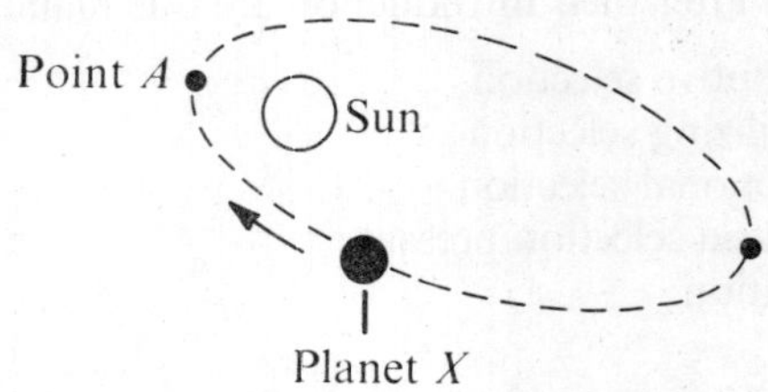

According to the laws of planetary motion, which of the statements below describes the motion of planet X as it moves in its orbit through points A and B in the diagram above?

(A) Its speed increases as it approaches point A and decreases as it approaches point B.
(B) Its speed decreases as it approaches point A and increases as it approaches point B
(C) Its speed remains unchanged as it approaches point A but increases as it approaches point B.
(D) Its speed remains unchanged as it approaches point A but decreases as it approaches point B.
(E) Its speed is constant through all points on its orbit.

Johannes Kepler (1571-1630) formulated three laws of planetary motion using the data of Tycho Brahe (1546-1601). These laws stated that (1) all planets move in elliptical orbits having the Sun as one focus of the ellipse; (2) an imaginary line from a planet to the Sun sweeps out equal areas in the same amount of time; and (3) $r^3/T^2 = k$, where r is the planet's mean distance (radius) from the Sun, T is the time the planet takes to travel once around the Sun, and k is a constant. The five choices relate to Kepler's second law. For equal areas to be covered in the same time, the planet X must travel faster as it approaches point A and then slow down as it approaches point B. Thus, choice A is the correct answer.

---

68. Bioengineering through new genetic technology is revolutionizing modern medical testing procedures. One new method designed to test for the presence of a defective allele even when the exact nucleotide sequence change is not known is called

(A) karyotyping of humans
(B) ultrasound fetal monitoring
(C) restriction-fragment-length polymorphism (RFLP) testing
(D) amniocentesis
(E) chorionic villi sampling

In order to test for the presence of a defective allele, chromosomes must be extracted from cells and various tests performed on them. Karyotyping refers to the visualization of metaphase chromosomes in a sequence based on chromosomal length, the position of the centromere, and banding. Although gross abnormalities can be seen, it is not possible to see individual alleles. Ultrasound fetal monitoring is used to view the fetus in the uterus. Amniocentesis involves the removal of fluid from around the fetus. Cells in the fluid are cultured and then the cells are analyzed for chromosomal and genetic defects using a variety of techniques. Chorionic villi sampling takes cells from the fetal membrane and uses them to look for chromosomal abnormalities. Restriction-fragment-length polymorphism testing utilizes the DNA from cells collected during amniocentesis, chorionic villi sampling, or other tissue sampling. The DNA is digested using a restriction enzyme. The presence or absence of a defective allele can then be determined by using radioactive probes. Choice C, therefore, is the correct answer.

---

69. It is known that heterozygous individuals living in central Africa who carry a gene for normal hemoglobin and a gene for sickle cell anemia are resistant to malaria. Whenever a heterozygous condition confers a selective survival advantage over homozygous types, this situation results in a phenomenon called

(A) genetic drift
(B) gene migration
(C) balanced polymorphism
(D) differential reproduction
(E) differential fecundity

Susceptibility to malaria is much lower in individuals who are heterozygous for sickle cell. Children who do not carry a gene for sickle cell have a much higher chance of contracting malaria and dying. Similarly, individuals homozygous for sickle cell usually die at an early age. Thus, the heterozygous condition is advantageous. Polymorphism is defined as the presence within a population of two or more phenotypically distinct forms of a trait. In the question, sickle cell is the trait. The polymorphism is balanced because the sickle-cell allele is maintained at high frequencies due to its selective advantage. Thus, choice C is the correct answer.

---

70. Which of the following enzymes would be best to use for cleaving a molecule of double-stranded DNA into segments of various sizes?

(A) Restriction enzyme
(B) DNA polymerase
(C) Exonuclease
(D) Reverse transcriptase
(E) DNA ligase

All five enzymes are concerned with the processing of nucleic acid. Three of them are used in synthesis: DNA polymerase, reverse transcriptase, and DNA ligase; and two are used in breakdown: restriction enzyme and exonuclease. The question asks for an enzyme that cleaves DNA into segments of various sizes. The exonuclease cleaves nucleotides one at a time from the end of a polypeptide chain, but the restriction enzyme cleaves within the DNA chain and can therefore be used to break the DNA into segments of various sizes. Thus, choice A is correct.

---

71. Eggs with shells evolved first in which of the following groups?

(A) Fishes
(B) Amphibians
(C) Birds
(D) Mammals
(E) Reptiles

Vertebrates became fully adapted to land with the development of the amniotic egg. This egg retained its own water supply, yolk, and albumin and was surrounded by a shell and an egg membrane, both of which were waterproof but allowed for diffusion of gases. Such eggs were first found in the reptiles; thus, choice E is the correct answer.

---

72. All of the atoms of sodium, whether or not in compounds, must have exactly eleven of which of the following?

(A) ions
(B) protons
(C) neutrons
(D) electrons
(E) photons

Atoms are made up of nuclei that contain protons and neutrons, which can be further subdivided into subatomic particles. Electrons orbit the nuclei. Atoms differ from one another in their number of protons, neutrons, and electrons. It is the number of protons that determines the type of atom. Sodium atoms always contain eleven protons. Thus, choice B is the correct answer.

---

73. School policy statements concerning the handling of students who develop AIDS can be best used by classroom teachers to

(A) introduce a unit on bacteria
(B) teach a lesson on DNA
(C) develop concepts about laws pertaining to genetic screening
(D) show how scientific information relates to decisions made in society
(E) explain how a vaccine is made

The presence of the human immunodeficiency virus (HIV) may lead to the development of AIDS. Classroom teachers can use school policy statements concerning students who develop AIDS in a discussion on societal decisions and how these decisions relate to scientific evidence. The statements would not be relevant to bacteria, because AIDS results from the presence of a virus. Similarly, while some viruses contain DNA, HIV does not; it is a retrovirus and contains RNA. Genetic screening is done to ascertain whether or not individuals contain certain alleles that may lead to disorders such as phenylketonuria. This does not pertain to AIDS, which is not a hereditary disease. Vaccines are produced to help prevent viral and bacterial diseases. Explaining how one is produced bears no relation to school policy statements concerning the handling of students who develop AIDS. The best answer, therefore, is choice D.

---

74. During metaphase in the process of mitosis, the line-up of chromosomes along the equator helps ensure that

(A) the newly formed daughter cells contain a copy of each chromosome
(B) the homologous chromosomes stay within the nucleus of the cell
(C) the chromatid pairs remain together
(D) gametes are formed
(E) crossing-over occurs

Mitosis is the process in which chromosomes previously replicated are divided into two identical nuclei. In order to make certain that the DNA in both nuclei is identical, the chromosomes follow a defined pathway. When they line up at the equator during metaphase, the replicated sister chromatids remain attached at a point called the centromere. The sister chromatids then separate forming two chromosomes, one of which travels to one pole and the other to the second pole. This ensures that the newly formed daughter nuclei contain a copy of each chromosome. Choice A is the correct answer.

---

75. The cellular synthesis of RNA from a DNA template is called

(A) duplication
(B) translation
(C) transcription
(D) transformation
(E) replication

Duplication and replication have very similar meanings, that is, to copy. The duplication of DNA is known as replication and this is not being asked. Choices B and D are incorrect because translation refers to the assembly of amino acids into a protein based on the sequence of bases on messenger RNA, and transformation takes place when DNA is taken into a cell from the external medium. Since transcription is the process of making a complementary strand of mRNA from a DNA template, choice C is the correct answer.

---

76. A population curve shows a rapid increase initially, then levels off to a rather stable number of individuals. At this stable level, the population has reached

(A) the biotic potential
(B) the carrying capacity
(C) the saturation point
(D) a free-running period
(E) biological magnification

After initial rapid growth, a model population starts to slow down until it reaches a steady state. At this stable level, the population that exists is the maximum the environment can support and is known as the carrying capacity. Thus, choice B is the correct answer. The biotic potential refers to the maximal rate of increase in number of individuals of the population and is not at a "stable level." The saturation point is a point at which a solvent cannot hold any more solute and is not used when describing logistic growth curves. A free-running period refers to deviations from normal circadian rhythms and thus is not relevant. Biological magnification is a trophic process. Toxins become more concentrated in the tissues of an organism the further along a food chain it is. This also has nothing to do with population growth curves.

---

77. Which of the following local changes in atmospheric conditions would most likely result in rain?

(A) A decrease in relative humidity
(B) A decrease in atmospheric pressure
(C) An increase in carbon dioxide concentration
(D) An increase in air density
(E) A replacement of cool, humid air by warm, dry air

A decrease in relative humidity means that the air becomes less saturated with water and would not normally lead to rain. An increase in carbon dioxide concentration also would have no effect on the potential for rain since it is the amount of water in the air coupled with temperature that is important. The greater the concentration of water vapor in the air the less dense it is; thus, by increasing the air density, less water would be present and rain would not be likely to occur. A replacement of cool, humid air with warm, dry air would mean a decrease in humidity, and rain would not be likely to occur in this situation. A decrease in atmospheric pressure would likely be accompanied by an increase in water vapor concentration. If the air mass is cold enough, the water molecules will combine to form raindrops and rain will fall. Thus, choice B is the correct answer.

---

78. $\ldots C_8H_{18} + \ldots O_2 \rightarrow \ldots CO_2 + \ldots H_2O$

When the equation above is balanced with the lowest whole-number coefficients, what is the coefficient for CO_2?

(A) 4
(B) 8
(C) 12
(D) 16
(E) 18

$$2\,C_8H_{18} + 25\,O_2 \rightarrow 16\,CO_2 + 18\,H_2O$$

In a balanced equation the number of atoms of each element must be equal on both the left-hand side (reactant side) and the right-hand side (product side) of the equation. In the balanced equation above, there are 16 carbon atoms on the left-hand side of the equation, and therefore there should also be 16 on the right-hand side. Thus, choice D is the correct answer.

---

79. If a series circuit powered by a 6-volt battery has a current of 2 amperes, the total resistance in that circuit is

(A) 3 ohms
(B) 4 ohms
(C) 6 ohms
(D) 8 ohms
(E) 12 ohms

Ohm's law states that V/I = R, where V is the voltage (volts), I is the current (amps), and R is the resistance (ohms). V = 6 volts, I = 2 amps, and R = 6 volts/2 amps = 3 ohms. Thus, choice A is the correct answer.

---

80. Which of the following hypotheses is LEAST amenable to verification by experimentation?

(A) The average size of eukaryotic genes is greater than the average size of prokaryotic genes.
(B) *Archaeopteryx* evolved from a reptilian ancestor.
(C) Triploid watermelons exhibit defective meiosis and are therefore seedless.
(D) The DNA that codes for the beta chain of hemoglobin has more introns than the DNA that codes for the alpha chain.
(E) Humans and chimpanzees share some of the same genes.

Choices A, C, D, and E all involve experimentation in the areas of cell biology and molecular biology. There are well-documented methods on how to carry out the experiments. With respect to choice B, it is very difficult to verify; *Archaeopteryx* is now extinct, as is its supposed reptilian ancestor. Thus, while evidence may be obtained to support the hypothesis, it is most unlikely that it could be verified. Choice B, therefore, is correct.

---

81. FOOD FROM 2.27 KILOGRAMS OF GRAIN

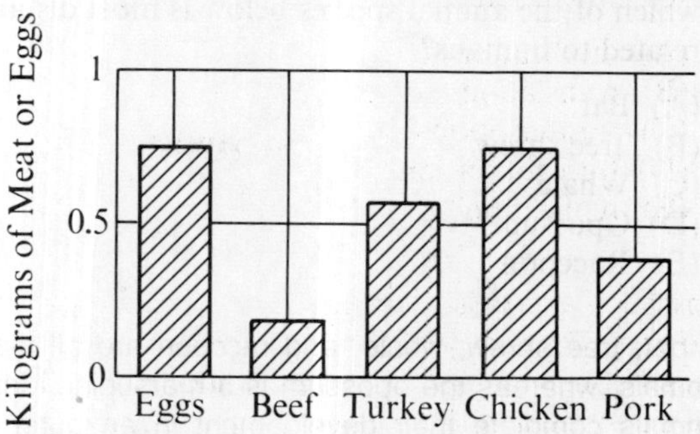

The graph above illustrates the numbers of kilograms of meat or eggs produced for every 2.27 kilograms of grain eaten by an animal. Which of the following changes in the eating habits of people in the developed world would result in the most efficient use of grain?

(A) Eating more pork than eggs
(B) Eating more chicken than beef
(C) Eating more beef than eggs
(D) Eating more turkey than chicken
(E) Eating more eggs than chicken

The bar graph shows that 2.27 kilograms of grain produces more eggs and more chicken than either beef, turkey, or pork, and that it produces the least amount of beef. A change in eating habits that would result in the most efficient use of grain would be to eat more chicken (or eggs) than beef. The correct answer is choice B.

---

82. According to the theory of plate tectonics, which of the following statements is correct?

(A) Crustal plates are destroyed at the midocean ridges.
(B) The Earth's crust is one solid plate made of basalt.
(C) The outer core of the Earth is made of solid material.
(D) Earthquakes and volcanoes usually occur in the middle of crustal plates.
(E) Continents drift because they are located on crustal plates that are moving.

According to the theory of plate tectonics, the Earth's crust can be divided into six large crustal plates. Crustal plates can be destroyed, although not at the midocean ridges where new crust is formed. The Earth's outer crust is not a single solid plate. The outer core of the Earth is believed to be liquid, not solid. Earthquakes and volcanoes usually occur at the edges of the crustal plates and not in the middle. Crustal plates do move and continents are carried by them; therefore, choice E is correct.

---

83. From the perspective of its evolutionary origins, which of the animal species below is most distantly related to humans?

(A) Bat
(B) Tree shrew
(C) Whale
(D) Opossum
(E) Raccoon

The bat, tree shrew, whale, and raccoon are all placental mammals, whereas the opossum is a marsupial. Marsupial mammals complete their development in an outer pouch called a marsupium. Evidence suggests that the marsupials and placental mammals diverged from a common ancestor about 80 to 100 million years ago. Thus, choice D is the correct answer.

---

84. All of the following are likely to result from cutting down large areas of tropical rain forests EXCEPT

(A) an increase in the species diversity of the areas
(B) an increase in erosion by rivers flowing through the areas
(C) an increase in the rate of nutrient loss from the areas
(D) an increase in the percentage of carbon dioxide in the air
(E) a decrease in the percentage of oxygen in the air

When large areas of tropical rain forests are cleared, extensive erosion by rivers flowing through the areas increases due to the reduction of extensive root systems. This reduction then leads to an increase in the rate of nutrient loss from the areas. The percentage of carbon dioxide in the air will increase because there will be fewer plants photosynthesizing, and, similarly, there will be a decrease in the percentage of oxygen in the air because oxygen is given off during photosynthesis. Choice A is the exception. It is thought that the continuing loss of the incredibly diverse habitats found in tropical rain forests will cause the extinction of approximately one million species by the beginning of the next century. Thus, choice A is the correct answer.

---

85. Each of the several species of *Yucca* found in the North American desert can only be pollinated by a unique species of yucca moth. Each species of moth is dependent on its species of plant for food and reproduction. This relationship is best described as

(A) parasitism
(B) endosymbiosis
(C) mutualism
(D) commensalism
(E) predation

A relationship in which one organism is harmed and the other benefits is called parasitism. If one organism (a prokaryote or a eukaryote) lives inside the cells of another eukaryote, the whole behaving typically as an organism, it is known as endosymbiosis. If one organism is helped, but the other is not affected, it is called commensalism, and if both organisms benefit, it is called mutualism. Predation refers to one organism preying on another organism, and this is not described in the question. The *Yucca* plant and yucca moth both benefit from the relationship because the *Yucca* plant is pollinated and the yucca moth obtains food. Thus, the correct answer is mutualism, choice C.

---

86. Which of the following elements makes up the greatest percentage by mass of the Earth's crust?

(A) Iron
(B) Carbon
(C) Nitrogen
(D) Silicon
(E) Sulfur

The eight most common elements in the Earth's crust in order of percent by mass are oxygen (46.6), silicon (27.7), aluminum (8.1), iron (5.0), calcium (3.6), sodium (2.8), potassium (2.6), and magnesium (2.1). Of the choices given, silicon, choice D, is clearly the most abundant.

---

87. The presence of which of the following is useful in separating ferns and mosses into different taxonomic categories?

(A) Chlorophyll
(B) Vascular tissue
(C) Nuclei
(D) Chloroplasts
(E) Flagellated sperm

Ferns and mosses are found in the Kingdom Plantae. Chlorophyll, nuclei, and chloroplasts cannot be used to separate ferns and mosses into different taxonomic categories as these characteristics are common to all organisms in the Kingdom Plantae. Flagellated sperm are characteristic of plants that rely on water for fertilization. Both mosses and ferns have flagellated sperm. Mosses are members of the division Bryophyta and ferns of the division Pterophyta. The Bryophytes lack specialized tissue (vascular tissue) to carry water and nutrients, namely, the xylem and phloem, whereas the Pterophytes contain these specialized tissues. Thus, ferns and mosses can be separated into two different taxonomic groups based on the presence or absence of the conducting vascular tissue. Choice B, therefore, is the correct answer.

---

88. When moose migrate, the weaker ones often become the prey of wolves and other carnivores. If, in a particular year, vegetation eaten by moose was sparse in an area, which of the following statements is most likely correct concerning the wolf population?

(A) The wolf population will not change.
(B) The wolf population will initially increase due to an increase in food.
(C) The wolf population will initially decrease because moose would become stronger by fighting over the food.
(D) The wolf population will increase simultaneously with the moose population.
(E) The gain in wolf biomass will be equal to the quantity of the moose eaten.

If there were a shortage of food for the moose, many would become weak. This weakening would make more of them vulnerable to the wolves. Those wolves that would in a normal year die may, because of the excess food available, survive and reproduce. Thus, the wolf population will initially increase due to an increase in food, and choice B is the best answer.

89.

I. Fish skeleton
II. Rodent skull
III. Trilobite
IV. Cyanobacteria
V. Salamander-like creature

Students were presented with the representative fossils listed above. They were asked to place the fossils in the order in which their respective groups evolved. Starting with the oldest group, which of the following sequences is correct?

(A) I, II, III, IV, V
(B) II, V, I, III, IV
(C) III, IV, II, V, I
(D) IV, III, I, V, II
(E) IV, III, V, I, II

Of the five representative fossils given (fish, rodent, trilobite, cyanobacteria, and a salamander-like creature), the cyanobacteria are the oldest organisms in evolutionary terms, followed by the trilobites, which were first present in the Devonian period (360 to 400 million years ago). Fish evolved next, followed by the salamander-like creature, an amphibian. Mammals are believed to have appeared last. The sequence from oldest group to youngest group, therefore, is choice D, cyanobacteria (IV), trilobite (III), fish skeleton (I), salamander-like creature (V), and rodent skull (II).

---

90. A student presents a current-events report on using engineered microorganisms to clean up specific hazardous-waste sites. Of the following questions posed by other students, which has the LEAST direct relevance to the topic of the report?

(A) How can microorganisms be engineered?
(B) Why can these microorganisms digest the wastes and other microorganisms cannot?
(C) What happens to the wastes?
(D) How do microorganisms cause disease in humans?
(E) How can these microorganisms fit into the scheme of other decomposers in the area?

Choices A, B, C, and E all list questions that are directly related to the topic of using engineered microorganisms to clean up hazardous-waste sites. The question given as choice D, on the other hand, has nothing to do directly with the use of engineered microorganisms to clean up hazardous-waste sites. Choice D is least relevant to the topic and is, therefore, the correct answer.

---

91. Which of the following isolating mechanisms is generally considered to initiate the process of speciation in the majority of cases?

(A) Mechanical isolation
(B) Behavioral isolation
(C) Developmental isolation
(D) Seasonal isolation
(E) Geographic isolation

Speciation is the process by which new species are formed. It arises through reproductive isolation. It is currently thought to be due mostly to the geographic isolation of a population of organisms. Thus, choice E is the correct answer. Although the other four types of isolation may separate populations for a time and lead to some speciation, they are not thought to initiate the process of speciation in the majority of cases.

---

92. A segment of DNA that codes for a polypeptide containing 120 amino acids must have a minimum of how many nucleotide base pairs?

(A) 40
(B) 120
(C) 240
(D) 360
(E) 480

Three adjacent nucleotides in a molecule of DNA code for a specific amino acid or for termination of a polypeptide chain. There are 120 amino acids in the polypeptide in the question. If each amino acid is coded for by 3 bases, then 120 times 3 nucleotide base pairs are the minimum (360 base pairs) needed to code for a polypeptide containing 120 amino acids. Thus, choice D is the correct answer.

---

93. On a lush 35-square-mile island, fifty miles from the nearest coast of the mainland, 5 male deer and 17 female deer were introduced. No predators of deer were on the island and no hunting was allowed. The graph below shows the population of deer over 50 years.

DEER POPULATION GRAPH

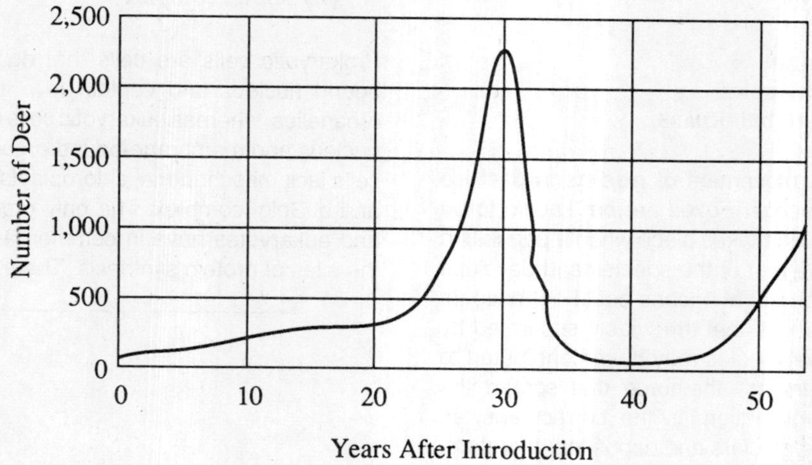

Statements that are probably correct about the deer population include which of the following?

I. The deer multiplied because of the abundance of food.
II. The deer reached high population density because of a lack of predators.
III. The deer population could have increased indefinitely if it were not for diseases that probably spread through the herd.

(A) I only
(B) II only
(C) III only
(D) I and II only
(E) II and III only

The graph shows the deer population increasing slowly at first and then peaking rapidly after 30 years of inhabiting the island. Within a few years, the total fell from a peak of about 2,250 deer until it numbered less than 100. The population then began to increase in number again. Statement I is probably correct because the question indicates that the island was lush. Statement II is probably also correct, as no hunting was allowed on the island and there were no predators there; these factors, coupled with the abundant food, meant that the growth of the deer could continue unabated. Statement III is clearly incorrect because, at some point, food, space, or some other factor would become limiting. Thus, choice D, "I and II only," is the correct answer.

94. Several plant species growing on moraines of lower New York State and New Jersey are known to grow otherwise only in northern Canada. Which of the following mechanisms is most likely responsible for the existence of these plant species in the United States?

(A) Migration of foxes
(B) Glaciation
(C) Divergent evolution
(D) Shifts in magnetic poles
(E) Rising levels of global oceans

This question relates the movement of glaciers in past ice ages to the dispersal of seeds. Foxes are not known to be migratory. Divergent evolution takes place when a population becomes isolated from the rest of the species and develops differently. Rising levels of global oceans could not account for the two similar populations that are widely separated by landmass. Shifts in magnetic poles may have contributed to glaciation, but the glaciers are the force that spread the seeds. Thus, choice B, glaciation, is the correct answer. Seeds were carried by the glaciers and deposited; therefore, the same plant species are found growing on moraines in lower New York State and New Jersey, as well as in northern Canada.

95. Which of the following can be found in both prokaryotic cells and eukaryotic plant cells?

(A) Mitochondria
(B) Chloroplasts
(C) Ribosomes
(D) Endoplasmic reticulum
(E) Golgi complex

Prokaryotic cells are cells that do not have a membrane-bound nucleus and contain few, if any, membrane-bound organelles, whereas eukaryotic cells have a membrane-bound nucleus and membrane-bound organelles. Thus, prokaryotic cells lack mitochondria, chloroplasts, endoplasmic reticulum, and a Golgi complex. The only organelles that prokaryotes and eukaryotes have in common are ribosomes, which are the sites of protein synthesis. Therefore, choice C is correct.

Questions 96-98

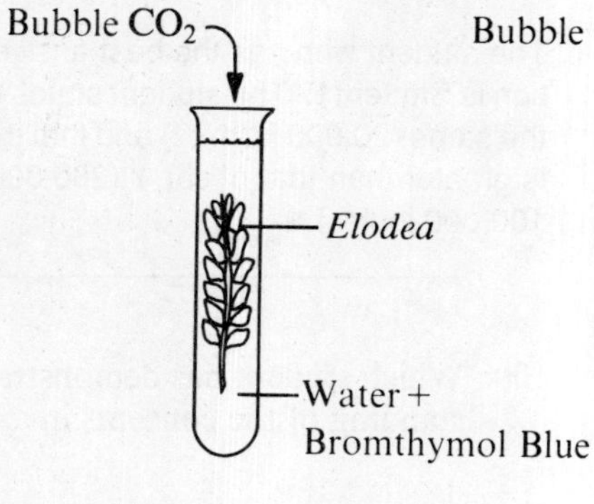

Figure 1 - Experimental

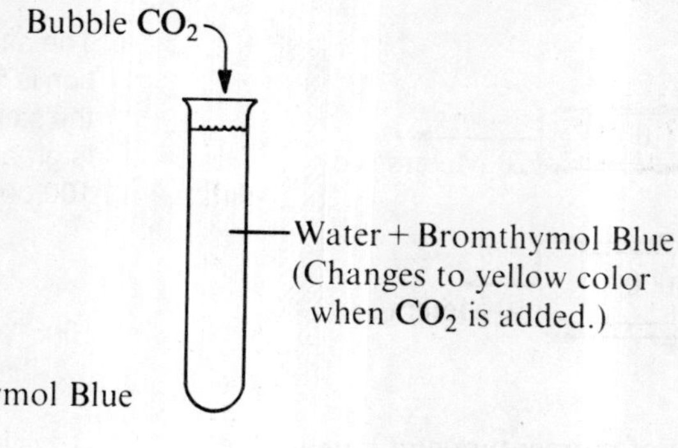

Figure 2 - Control

A biology teacher sets up an experiment as indicated in the diagrams above. A sprig of *Elodea*, water, and bromthymol blue indicator are placed in a test tube. Carbon dioxide is bubbled through the solution for a few minutes, which changes the color from blue to yellow. The plant is allowed to photosynthesize. During the day, the yellow solution turns blue; at night, it turns back to yellow. The control for this demonstration is a test tube of water and bromthymol blue, which turns yellow when carbon dioxide is bubbled through the solution.

The students are asked to explain what caused the color changes observed in the experimental tube shown in Figure 1. Five students respond as follows.

Student A: Bromthymol blue turns yellow in an acid medium and blue in a basic medium.

Student B: Photosynthesis uses up carbon dioxide, causing the solution to turn yellow, and cellular respiration causes the solution to turn blue.

Student C: Carbon dioxide, water, sunlight, and chlorophyll are all necessary for photosynthesis.

Student D: Oxygen is released during photosynthesis, which causes the solution to turn blue, then yellow at night.

Student E: During the day, photosynthesis uses CO_2 and causes the yellow color to change to blue; at night, the release of CO_2 by cellular respiration causes the blue color to change to yellow.

In acid solution, bromthymol blue is yellow; in alkaline solution, bromthymol blue is blue. As the *Elodea* photosynthesized during the day, carbon dioxide was removed from the water. This made the water alkaline; hence the indicator, bromthymol blue, changed from yellow to blue. At night, when carbon dioxide was given off due to respiration, the CO_2 combined with water to form carbonic acid and the bromthymol blue turned from blue to yellow. The students were asked to explain the color changes.

96. Which student understands the requirements of photosynthesis, but has not answered the question?

The student who knew what the raw materials of photosynthesis are was Student C. The student did not, however, say why the bromthymol blue changed color and thus did not answer the question.

97. Which student best answers the question?

Student E knew that the plant used CO_2 to photosynthesize and that this caused the bromthymol blue to turn blue. The student also knew that CO_2 is released during respiration and that the release would cause the bromthymol blue to turn yellow again. Thus, Student E best answered the question.

98. Which student best understands the chemical reasons for the color change, but did not answer the question?

Student A knew that bromthymol blue turned yellow in acid medium and blue in basic medium, but did not explain why. Thus, Student A best understood the chemical reason for the color change, but did not answer the question.

Questions 99-102

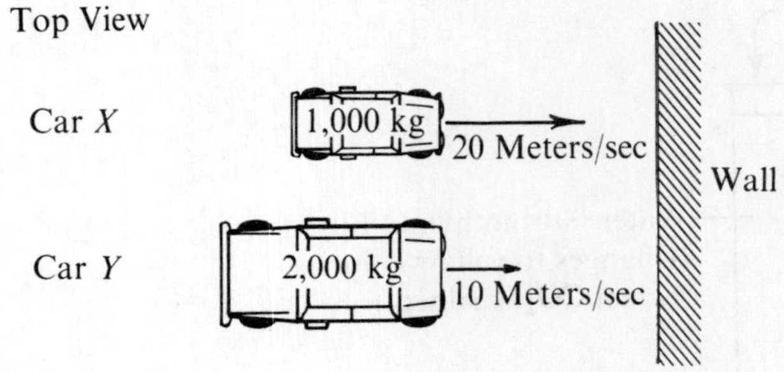

During a discussion of momentum and kinetic energy, the teacher draws on the board the diagram above, and asks the class to compare the momentums and the kinetic energies of the two cars. Five students respond as follows:

Student A: Car *X* is lighter and so both its momentum and its kinetic energy are less than the values for car *Y* are.

Student B: There is too little information for me to make any comparisons.

Student C: The kinetic energies are the same, but car *Y* has greater momentum since it is heavier.

Student D: Since both momentum and kinetic energy include both a mass term and a speed term, and since car *X* has twice the speed of car *Y* but car *Y* weighs twice as much as car *X*, the kinetic energy and the momentum for the two cars are the same.

Student E: The momentums are the same, but car *X* has the greater kinetic energy since its speed is greater and kinetic energy has a speed squared term.

Momentum is the product of the mass of an object and its velocity; i.e., momentum = (mass) (velocity) [momentum = mv]. Kinetic energy is the energy of an object due to its motion and is equal to one-half the product of the mass and the velocity squared ($KE = 1/2\ mv^2$).

The teacher asks the students to compare the momentums and the kinetic energies of the two cars.

Car *X* has a momentum of $1{,}000\ kg \times 20\ m{\cdot}s^{-1} = 20{,}000\ kg{\cdot}m{\cdot}s^{-1}$.
Car *Y* has a momentum of $2{,}000\ kg \times 10\ m{\cdot}s^{-1} = 20{,}000\ kg{\cdot}m{\cdot}s^{-1}$.
Car *X* has a kinetic energy of $1/2 \times 1{,}000\ kg \times (20\ m{\cdot}s^{-1})^2$
= 200,000 joules.
Car *Y* has a kinetic energy of $1/2 \times 2{,}000\ kg \times (10\ m{\cdot}s^{-1})^2$.
= 100,000 joules.

99. Which student has the best answer to the teacher's question?

The student who has the best answer to the teacher's question is Student E. The student states that the momentums are the same ($20{,}000\ kg{\cdot}m{\cdot}s^{-1}$) and that the kinetic energy of car *X* is greater than that of car *Y* (200,000 joules as compared to 100,000 joules).

---

100. Which student has demonstrated the LEAST understanding of the concepts involved?

There is enough information to answer the teacher's question. Student B obviously has not understood the principles of momentum or kinetic energy.

---

101. Which student has shown enough understanding that seeing the equation defining kinetic energy should lead the student to correct the error?

Student D has not realized that kinetic energy has a squared term in it. If the student saw the equation, the student would realize that the kinetic energies of the two cars differ.

---

102. Which student has given an answer that would be correct if the two cars were moving at the same speeds?

If the two cars were moving at the same speed (e.g., 10 m/s), the momentum of car *X* would be $10{,}000\ kg{\cdot}m{\cdot}s^{-1}$ and of car *Y*, $20{,}000\ kg{\cdot}m{\cdot}s^{-1}$, and the kinetic energy of car *X* would be 50,000 joules and of car *Y* would be 100,000 joules. Student A states that car *X* is lighter and so both its momentum and kinetic energy are less than the corresponding values for car *Y*. This would be true if the two cars moved at the same speed.

Questions 103-104

(A) Leech
(B) Mosquito
(C) Lichen
(D) Fungus
(E) Flea

103. Is a parasitic annelid

Of the five choices (leech, mosquito, lichen, fungus, and flea), only the leech is an annelid. Leeches are primarily bloodsuckers with digestive tracts that are especially adapted for storing blood. Choice A, therefore, is the correct answer.

104. Is an example of a mutualistic relationship

A mutualistic relationship takes place when different species live together and each benefits the other. A lichen is a combination of a fungus and a blue-green alga or a green alga. Each organism supplies the other with certain nutrients. Therefore, choice C is the correct answer.

Questions 105-106

(A) Anton van Leeuwenhoek
(B) Linus Pauling
(C) Louis Pasteur
(D) Stanley Miller
(E) Carolus Linnaeus

105. Provided the system of binomial nomenclature currently used to classify plants and animals

The binomial system of nomenclature was put forward by the Swedish botanist Carolus Linnaeus. Thus, choice E is the correct answer.

106. Performed an experiment that demonstrated amino acid synthesis in a simulated primitive atmosphere

The person who demonstrated in the laboratory that amino acids could be made from a simulated primitive atmosphere was Stanley Miller, choice D.

Questions 107-108

These questions ask how organisms are classified.

(A) Arthropoda
(B) Annelida
(C) Mollusca
(D) Echinodermata
(E) Cnidaria (Coelenterata)

107. Hydra, jellyfish, sea anemone

Hydra, jellyfish, and sea anemones are all found in the phylum Cnidaria, choice E. They are made up of two tissue layers called ectoderm and endoderm that are separated by an acellular mesoglea. These tissues surround a gastrovascular cavity. The ectoderm contains specialized cells called cnidocytes that are used for catching prey. The cnidocytes are found on tentacles surrounding the mouth.

108. Lobster, spider, tick

Lobsters, spiders, and ticks are all in the phylum Arthropoda, choice A. These organisms are segmented and have jointed appendages. Each also has an exoskeleton made of chitin.

Questions 109-111

These questions relate to matching specific substances with their group of compounds.

(A) Protein
(B) Carbohydrate
(C) Alcohol
(D) Nucleic acid
(E) Steroid

109. Cholesterol

Cholesterol, which is found in the cell membranes of animals, is a steroid, choice E.

110. DNA

DNA is the genetic material that makes up chromosomes. It is a nucleic acid, choice D.

111. Glycogen

Glycogen is one of the primary food storage substances of most animals and fungi. It is a carbohydrate, choice B.

Questions 112-115

(A) Infrared rays
(B) Ultraviolet rays
(C) Radio waves
(D) Visible rays
(E) X-rays

All five choices are part of the electromagnetic spectrum, namely, infrared rays, ultraviolet rays, radio waves, visible rays, and x-rays.

112. Are detected by large antenna arrays and provide detailed images of astronomical objects such as quasars

To detect long wavelengths such as radio waves, choice C, large antenna arrays are needed to measure the intensity of emission and establish its spectrum. Image analysis can then be used to provide an image similar to a photograph.

113. Are associated with vibrations in molecules and generally perceived as heat

Infrared rays, choice A, are associated with vibrations in molecules and are generally perceived as heat.

114. Primarily account for the perception of color as a result of selective absorption or reflection

Pigments in the human eye are sensitive to wavelengths present in the visible spectrum, choice D. Different wavelengths in this spectrum stimulate different combinations of the pigments, resulting in the perception of different colors.

115. Are associated with beta-particle emission by radioactive nuclides and with ejection of atomic K-shell electrons

In general, x-rays, choice E, are generated as the result of energy transitions of atomic electrons. Electrons are usually ejected from an inner shell and replaced by electrons from a higher energy level.

Questions 116-118

(A) Osmosis
(B) Transpiration
(C) Pinocytosis
(D) Phagocytosis
(E) Translocation

Osmosis is the diffusion of water across a semipermeable membrane from a region of low solute concentration to a region of high solute concentration. Transpiration is the loss of water vapor from the stomata of plants. Pinocytosis is the process whereby liquids are taken into a cell via a vacuole that forms from the cell membrane. Phagocytosis is the process whereby solids are taken into a cell via a vacuole that forms from the cell membrane. Translocation is the transfer of the products of photosynthesis from one part of a plant to another.

116. The intake of fluid droplets by a cell

The intake of fluid droplets by a cell is pinocytosis, choice C.

117. The diffusion of water through a semipermeable membrane from a region of low solute concentration to a region of high solute concentration

The diffusion of water through a semipermeable membrane from a region of low solute concentration to a region of high solute concentration is osmosis, choice A.

118. The taking of solid materials into a cell by engulfment and the subsequent pinching off of the cell membrane

The taking of solid materials into a cell by engulfment and the subsequent pinching off of the cell membrane is phagocytosis, choice D.

Questions 119-121 refer to the diagram below of the cross section of a leaf.

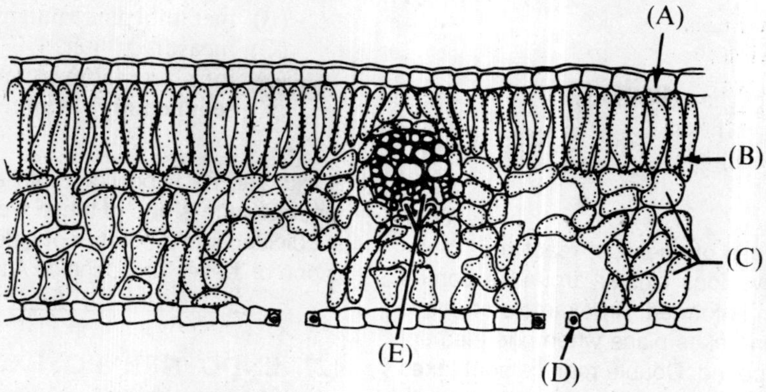

The cross section of leaf shows the epidermis (A), the palisade layer (B), the spongy mesophyll (C), the guard cell (D), and the phloem (E).

---

119. A site of greatest photosynthetic activity

A site of greatest photosynthetic activity is the palisade layer, choice B.

---

120. A site of translocation

A site of translocation is the phloem, choice E.

---

121. A site that regulates the amount of water lost from the leaf

A site that regulates the amount of water lost from the leaf is the guard cell, choice D.

Questions 122-125

(A) Complex ion formation
(B) Single replacement
(C) Double replacement (exchange of ions)
(D) Direct combination (synthesis)
(E) Decomposition (analysis)

All five choices refer to types of chemical reactions. When polar molecules or negative ions cluster around a central positive ion, a complex ion is formed. An example of this is $[NiBr_4]^{-2}$. Single replacement takes place when one element replaces another in a compound. Double replacement takes place when the positive part of one compound combines with the negative part of another compound and vice versa. Direct combination takes place when two or more elements or compounds combine. Decomposition or analysis is a reaction where a compound breaks down into two or more simple substances.

---

122. $CaO + H_2O \rightarrow Ca(OH)_2$

$CaO + H_2O \rightarrow Ca(OH)_2$
Direct combination, choice D, occurs when two compounds combine to form a third compound.

---

123. $2\ NaCl \rightarrow 2\ Na + Cl_2$

$2\ NaCl \rightarrow 2\ Na + Cl_2$
Decomposition, choice E, occurs when a compound decomposes into two or more substances.

---

124. $Cu + 2\ AgNO_3 \rightarrow 2\ Ag + Cu(NO_3)_2$

$Cu + 2\ AgNO_3 \rightarrow 2\ Ag + Cu(NO_3)_2$
Single replacement, choice B, occurs when the copper replaces the silver to make copper nitrate.

---

125. $AgNO_3 + NH_4Cl \rightarrow AgCl + NH_4NO_3$

$AgNO_3 + NH_4Cl \rightarrow AgCl + NH_4NO_3$
Double replacement, choice C, occurs when the silver ion and the ammonium ion change places.

---

126. IGNEOUS : ROCK ::
(A) potential : energy
(B) metamorphic : magma
(C) measured : liter
(D) infrared : radioactivity
(E) sedimentary : granite

Rocks can be subdivided into a number of different types. Igneous rock is one of these types. Similarly, energy can be subdivided into a number of different types. Potential energy is one of these. Thus, choice A is the correct answer.

---

127. ENDOTHERMIC : EXOTHERMIC ::
(A) acid : base
(B) refrigerator : furnace
(C) window : door
(D) cathode : anode
(E) proton : electron

The terms endothermic and exothermic are antonyms. An endothermic reaction is a chemical reaction that takes place with the absorption of heat, while an exothermic reaction is a chemical reaction that takes place with the evolution of heat. The best choice, therefore, is B, "refrigerator:furnace." A refrigerator is kept cold through heat exchange, whereas a furnace gives off heat. Thus, choice B is correct.

---

128. WIND SPEED : AIR PRESSURE GRADIENT ::
(A) erosion : deposition of sediment
(B) relative humidity : dew point
(C) river velocity : slope of riverbed
(D) air temperature : altitude
(E) heat : temperature

An air pressure gradient exists whenever air pressure changes from one place to another. When this occurs, the wind blows from regions where the air pressure is high toward low pressure areas. The steeper the gradient, the stronger the wind and vice versa. Choice C shows a similar pairing. As the slope of the riverbed becomes steeper, the velocity of the river will increase. The reverse also holds true. Thus, choice C is the correct answer.

---

129. ANODE : CATHODE ::
(A) rate : speed
(B) temperature change : volume change
(C) saturated solution : supersaturated solution
(D) pressure : contraction
(E) oxidation : reduction

A cathode is the electrode at which reduction occurs. An anode is the electrode at which oxidation takes place. Thus, choice E is the correct answer.

---

130. SCIENTIFIC METHOD : TRIAL AND ERROR ::
(A) pendulum : clock
(B) lever : fulcrum
(C) pump : handle
(D) professional : novice
(E) doctor : patient

A scientific method is a logical way of conducting research, whereas trial and error implies an opposite or ad hoc approach of conducting research. The correct choice is D. A professional would go about a task in a logical manner, while a novice, who is learning, may use more of a trial-and-error approach.

131. ADENINE : THYMINE ::
(A) adenine : guanine
(B) adenine : cytosine
(C) cytosine : uracil
(D) cytosine : thymine
(E) guanine : cytosine

Adenine and thymine are bases found in DNA that complement each other; i.e., if adenine is present in one strand of DNA, it is always joined to thymine in the other strand. Of the five choices given, the only other two bases that complement each other are guanine and cytosine. Thus, choice E is the correct answer.

132. VOYAGER : NEPTUNE ::
(A) oscilloscope : light
(B) electron microscope : mitochondria
(C) Bunsen burner : heat
(D) filter : water
(E) motor : boat

Voyager was a spacecraft used by the United States to travel past the planet Neptune and send information back about the planet. It was used to get a closer view of Neptune than can be had from Earth. The only option that involves looking at something more clearly is choice B. The electron microscope is used to view organelles such as mitochondria within cells. Photographs are then taken of what the organelles look like. Thus, choice B is correct.

Questions 133-137

The figure below shows a food web. Each letter represents a species. The arrows show the flow of materials and energy.

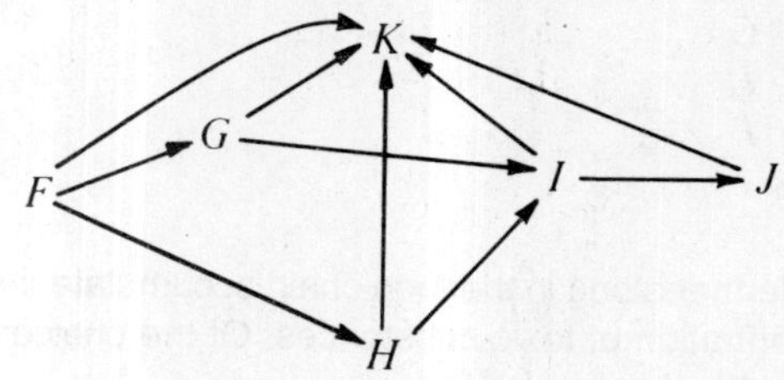

In the diagram of a food web, the direction of the arrows indicates the flow of materials and energy. Thus, *F* must be a producer (i.e., an organism such as a plant that produces its own food and serves as food for primary consumers). *G* and *H* must be primary consumers (herbivores) because they are consuming plants or algae. *I* is a secondary consumer (carnivore), and *J* is a tertiary consumer. *K*, which obtains food from all sources, is probably a detritivore (e.g., a decomposer).

133. The producer in this food web is

(A) F
(B) H
(C) I
(D) J
(E) K

The producer is *F*, choice A.

134. Which of the following would have the greatest total biomass?

(A) F
(B) J
(C) K alone
(D) $G + H$
(E) $I + K$

The total biomass is found at the producer level; thus *F*, choice A, is correct.

135. J is a

(A) primary consumer
(B) secondary consumer
(C) tertiary consumer
(D) decomposer
(E) primary producer

J is a tertiary consumer; therefore, choice C is correct.

136. Suppose that a toxic substance becomes concentrated in the food web by biological magnification. Of the following, the species likely to be most affected is

(A) *F*
(B) *G*
(C) *H*
(D) *I*
(E) *J*

Species farther along in the food chain accumulate the greatest concentration of toxic substances. Of the choices given, choice E, *J*, is the correct one.

---

137. Which of the following species is most likely the decomposer?

(A) *F*
(B) *H*
(C) *I*
(D) *J*
(E) *K*

The decomposer is *K*, choice E.

Questions 138-140

In a certain species of mammal, the allele for brown fur is dominant over the allele for white fur and the allele for long fur is dominant over that for short fur. An animal heterozygous for long, brown fur is mated to an animal that is homozygous for short, white fur. Assume that these alleles assort independently.

Let *L* = long fur. Let *B* = brown fur.
Let *l* = short fur. Let *b* = white fur.

Parents	*LlBb* x	*llbb*
Gametes	4 types *LB, Lb, lB, lb*	1 type *lb*

F_1 generation:
25 percent will have long, brown fur (*LlBb*),
25 percent will have short, brown fur (*llBb*),
25 percent will have long, white fur (*Llbb*),
and 25 percent will have short, white fur (*llbb*).

138. The percentage of the F_1 offspring expected to have white fur is

(A) 100%
(B) 75%
(C) 50%
(D) 25%
(E) 0%

The percentage of the F_1 offspring expected to have white fur (all *bb* offspring) is 50 percent, choice C.

---

139. The percentage of the F_1 offspring expected to be heterozygous for both fur color and fur length is

(A) 100%
(B) 75%
(C) 50%
(D) 25%
(E) 0%

The percentage of the F_1 offspring expected to be heterozygous for both fur color and fur length (all *LlBb* offspring) is 25 percent, choice D.

---

140. The percentage of the F_1 offspring expected to be homozygous for brown fur is

(A) 100%
(B) 75%
(C) 50%
(D) 25%
(E) 0%

There are no offspring with the genotype *BB*, and therefore the percentage of the F_1 offspring expected to be homozygous for brown fur is 0 percent, choice E.

Questions 141-143

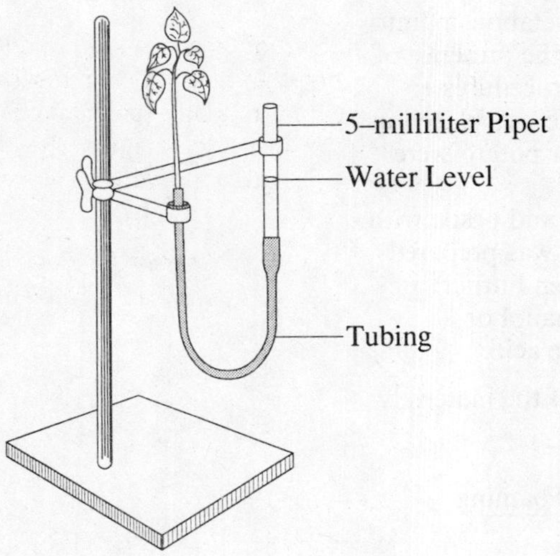

The stem of a plant shoot was held beneath water and cut diagonally. Tubing attached at one end to a pipet was filled with water and the tubing forced over the cut end of the shoot while still immersed in the water. The tubing and pipet were clamped as shown in the diagram above. Measurements were taken by reading the water level in the pipet every 15 minutes.

The stem of the plant was cut under water to make sure no air bubbles collected at the base. It was also inserted into the water-filled tubing under water so that air bubbles would not collect and affect the water flow.

---

141. The experimental setup is designed to measure

(A) root pressure
(B) photosynthesis
(C) transpiration
(D) guttation
(E) translocation

Because the question states that the water level in the pipet was measured every 15 minutes and since the setup was established under water to ensure a continuous flow of water, some form of water transport must be being measured. Root pressure is involved in water transport, but roots have to be present for it to be measured, and none are. Photosynthesis is still proceeding; however, it cannot be measured directly by using changing water levels. Guttation is the loss of liquid water through the leaves at their edges and is a visible consequence of root pressure. The setup cannot, therefore, measure this. Translocation is the process of transporting food substances, not water, around the plant, and therefore the setup cannot measure this either. Transpiration is the loss of water vapor from the plant. As the water molecules are lost through the stomata, they are replaced from the continuous column of water that extends from the roots to the leaves. As water molecules exit, more enter through the cut stem and the water level goes down in the pipet. Thus, loss of water (transpiration) is being measured, making choice C the correct answer.

---

142. Which of the following would have the LEAST effect on changing the water level from its initial value?

(A) Ambient temperature
(B) Number of leaves
(C) Humidity
(D) Dissolved CO_2 in the water
(E) Number of open stomata

Choices A, B, C, and E would affect the water level in the pipet. Ambient temperature will affect transpiration. The rate of evaporation increases with increasing temperature. Transpiration increases with increasing numbers of leaves and open stomata. Humidity also affects water loss. If the air is very dry, more water vapor will be lost. If the outside air is already laden with water, less water will exit via the stomata. Dissolved CO_2 in the water would affect the water level least since the proportion of water molecules to carbon dioxide molecules would still be extremely high, and carbon dioxide does not directly affect transpiration. Thus, choice D is the correct answer.

---

143. Which of the following actions is most likely to lead to the most rapid lowering of the water level in the pipet?

(A) Increasing the light intensity
(B) Sprinkling the leaves with water
(C) Covering the plant with a clear plastic bag
(D) Covering the leaves with petroleum jelly
(E) Covering the plant with a green plastic bag

For the water level to go down rapidly in the pipet, it would be expected that transpiration would proceed at a faster rate. Sprinkling the leaves with water, covering the plant with a clear plastic bag, and covering the plant with a green plastic bag will increase the humidity and limit transpiration. Petroleum jelly will block the stomata and limit transpiration. "Increasing the light intensity," choice A, will most likely stimulate the stomata to open and raise the surface temperature of the leaf. Transpiration will then increase, leading to the most rapid lowering of the water level. Therefore, choice A is the correct answer.

Questions 144-147

Catalase is an enzyme that facilitates the breakdown of hydrogen peroxide, a harmful product of cell metabolism, into water and oxygen. In an experiment to test for the presence of catalase in tissue, students used the following procedures.

Ten milliliters of 3 percent hydrogen peroxide was placed in a series of numbered tubes and pieces of liver or potato were dropped in using forceps.

The ground tissue was prepared in a mortar and pestle with some sand as an abrasive. The cooked material was prepared by boiling it in water in a test tube over a Bunsen burner. Liver was soaked for 24 hours in 70 percent ethanol or 0.1-molar lead nitrate or 0.1-molar hydrochloric acid.

The following data were collected for each of the materials tested.

Material	Foaming
Sand	–
Whole liver	+
Ground liver	++
Cooked liver	–
Whole potato	+
Ground potato	++
Cooked potato	–
Liver soaked in 70% ethanol	–
Liver soaked in 0.1-M lead nitrate solution	–
Liver soaked in 0.1-M hydrochloric acid	–

Key: (–) no foaming (+) foaming (++) much foaming

The experiment is designed to show the effects of various factors on enzyme activity. Catalase is the enzyme involved, and its presence can be detected by the fact that, when active, it will break down hydrogen peroxide to water and oxygen. The oxygen will be seen streaming to the surface (foaming). The factors tested were surface area, temperature, pH, and the presence of lead nitrate.

144. Ground tissue produces more foaming than whole tissue because

(A) lipids from the disrupted cell membranes absorb more hydrogen peroxide
(B) water from the disrupted vacuoles accelerates the chemical process
(C) more enzyme is released from the disrupted cells to react with the substrate
(D) lysosomes from the ground tissue are disrupted and release proteolytic enzymes
(E) mitochondria from the disrupted cells have greater access to molecular oxygen

More foaming was produced when the tissues were ground up than when they were whole. Catalase, which is present in the cells, was released during the grinding process and was able to react with the hydrogen peroxide more readily. More enzyme, therefore, is released from the disrupted cells to react with the substrate, and choice C is correct.

145. Which of the following methods is most convenient to test for the production of oxygen in this experiment?

(A) PTC paper
(B) Litmus paper
(C) Lime water
(D) Chromatography
(E) A glowing splint

PTC paper tests for the ability to taste PTC (phenylthiocarbamide). Litmus paper is used to detect acids and bases. Lime water turns milky in the presence of CO_2. Chromatography is used to separate components of a mixture. A glowing splint will relight when plunged into oxygen and thus is used to test for its presence. Choice E is correct.

146. Which of the following is the best explanation for the fact that the cooked tissue produced no foaming?

(A) The amino acid sequence in catalase has been changed.
(B) The cell membrane thickens with heating and prevents the release of enzymes.
(C) The messenger RNA has been destroyed and new proteins cannot be synthesized.
(D) The tertiary structure of catalase has been altered.
(E) The water in the cells has been evaporated and no chemical reaction can occur.

When the tissues were heated prior to their addition to hydrogen peroxide, no foaming resulted. Catalase was still present, but it had been deactivated. Its amino acid sequence, however, was not likely to have been changed. Choice A is therefore incorrect. Choice B is not a satisfactory explanation since the cell membrane does not thicken with heating and thus prevent the release of enzymes. Choice C does not address the question. The production of oxygen does not depend on the production of new catalase but on the amount of existing functional catalase that can come into contact with hydrogen peroxide. Choice E is clearly incorrect as the material was cooked by boiling in water. Choice D is the only feasible answer to the lack of enzyme activity shown by the cooked tissue. As the temperature rises, the movement of the atoms increases and the forces such as hydrogen bonds that hold the enzyme in its tertiary structure are disrupted. The polypeptide chains that make up the enzyme then become tangled and no longer function correctly. Thus, choice D is correct.

---

147. Based on the recorded data, which of the following experimental variables seems to have the LEAST effect on enzyme function?

(A) High temperature
(B) Lead nitrate
(C) Alcohol
(D) Tissue type
(E) Hydrochloric acid

As can be seen from the data, temperature, lead nitrate, alcohol, and hydrochloric acid all affected enzyme function as evidenced by lack of foaming. Both potato and liver produced foaming when whole or ground. Thus, the variable tissue type had the least effect on enzyme function and choice D is the correct answer.

Questions 148-151

A student in a laboratory is given 7 grams of iron filings and 4 grams of powdered sulfur. The two elements are thoroughly mixed together on a sheet of paper. (Molecular weights: iron = 55.85, sulfur = 32.07)

148. Which of the following is the best method for separating the iron from the mixture?

(A) Lifting out the iron particles by using a magnifying glass and tweezers
(B) Adding water and filtering
(C) Adding water and evaporating to dryness
(D) Heating the mixture
(E) Passing a magnet just above the mixture

Although the separation could be accomplished as described in choice A, it would be tedious and clearly not the best method among the choices given. Choices B and C both describe good methods for separating mixtures, provided one or more of the components dissolve in water; iron and sulfur are essentially insoluble in water. Since mixtures can be separated into their component parts only by using methods that involve physical changes, "heating the mixture," choice D, will not separate out the iron and sulfur; in fact, it will cause them to combine. "Passing a magnet just above the mixture," choice E, makes use of the property of magnetism that only the iron has. This property is made use of in recycling plants to separate iron and steel from nonmagnetic material such as paper. Thus, a magnet will collect the iron from the iron-sulfur mixture; choice E is the correct answer.

---

149. If the mixture of iron and sulfur is heated in a test tube, a red glow spreads throughout the tube. Heat energy is obviously released. After cooling, the test tube and its contents are weighed. If the empty test tube weighed 17 grams and if it is assumed that no atmospheric gases were involved in the reaction, the mass of the test tube and its contents should be approximately

(A) 6 grams
(B) 17 grams
(C) 27 grams
(D) 28 grams
(E) 29 grams

The law of conservation of mass states that the mass of the reactants must be equal to the mass of the products in a reaction. When 7 grams of iron filings are heated with 4 grams of sulfur and it is assumed that no atmospheric gases are involved, then approximately 11 grams of product should be formed. As the mass of the test tube is 17 grams, the total mass should be the sum of 11 grams and 17 grams. Thus choice D, "28 grams," is the correct answer.

---

150. If the test tube in which the iron and sulfur were heated is broken and the contents are mixed with carbon disulfide, none of the solid dissolves. Free sulfur is soluble in carbon disulfide. Conclusions supported by the information given include which of the following?

I. A chemical change has occurred.
II. The sulfur has completely reacted.
III. Iron is not soluble in carbon disulfide.

(A) I only
(B) I and II only
(C) I and III only
(D) II and III only
(E) I, II, and III

The solid that results after heating does not dissolve in carbon disulfide, a known solvent of sulfur; thus, the sulfur must be in a combined form. The information given only relates to sulfur, so no conclusions can be made regarding the iron. The conclusions that can be drawn are I, "A chemical change has occurred," and II, "The sulfur has completely reacted." The correct answer, therefore, is choice B, "I and II only."

---

151. The most probable name of the product formed by the reaction of iron filings with powdered sulfur is

(A) iron III (ferric) sulfate
(B) iron III (ferric) sulfite
(C) iron II (ferrous) sulfite
(D) iron II (ferrous) sulfide
(E) iron II (ferrous) oxide

As no atmospheric gases are involved in the reaction, only the elements iron and sulfur can be present in the product. The compounds in choices A, B, C, and E all contain iron, sulfur, and oxygen. The correct choice, therefore, is D, "iron II (ferrous) sulfide."

Questions 152-155 refer to the graph below, which depicts the oxygen-dissociation curves of the hemoglobin of various organisms.

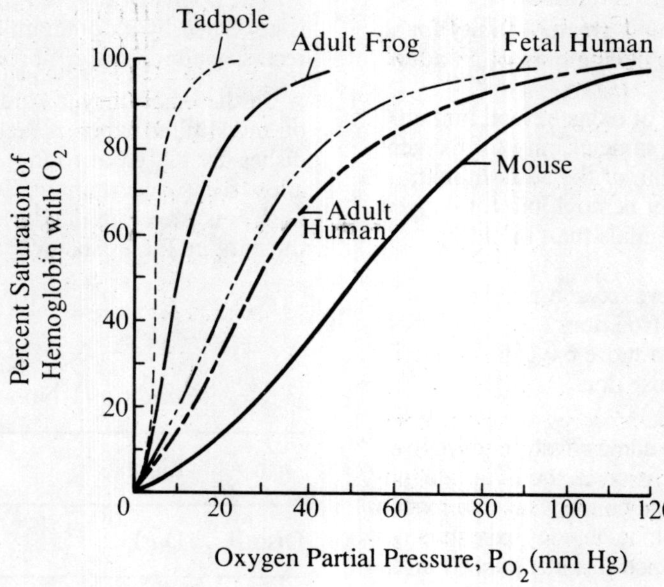

The graph shows how the amount of oxygen carried by hemoglobin is related to oxygen partial pressure in different animals.

152. If the P_{O_2} of mouse blood is 60 millimeters of mercury, what percentage of hemoglobin is saturated with oxygen?

(A) 10%
(B) 30%
(C) 60%
(D) 80%
(E) 100%

If an oxygen tension of 60 millimeters of mercury is followed to the oxygen-dissociation curve of a mouse, it can be seen by extrapolating from this curve to the ordinate that approximately 60 percent saturation is obtained. The correct answer, therefore, is choice C.

153. Which of the animals shows the greatest rate of change in oxygen uptake between 80 and 120 millimeters of mercury?

(A) Tadpole
(B) Adult frog
(C) Fetal human
(D) Adult human
(E) Mouse

The hemoglobin of the tadpole, adult frog, and fetal human is saturated with oxygen before or at 80 millimeters of mercury. The adult human increases saturation from 90 percent to 100 percent between 80 and 100 millimeters. The greatest rate of change in oxygen uptake between 80 and 120 millimeters of mercury is seen in the mouse, where the saturation of hemoglobin with oxygen goes from approximately 80 percent to 100 percent. Choice E, therefore, is the correct answer.

154. The oxygen-dissociation curves provide evidence that

(A) tadpole hemoglobin has a greater affinity for oxygen than has the hemoglobin of an adult frog
(B) at all partial pressures of oxygen, fetal human hemoglobin has the same affinity for oxygen as has the hemoglobin of the adult human
(C) the carrying capacity of hemoglobin for oxygen is less in younger animals than in older animals
(D) a tadpole lives in a more oxygen-rich environment than an adult frog does
(E) a fetal human lives in a more oxygen-rich environment than a mouse does

If fetal human hemoglobin has the same affinity for oxygen as does adult human hemoglobin, choice B, then the fetal human and adult human curves would be identical. Information about the carrying capacity of hemoglobin for oxygen is not presented in the graph, hence choice C is incorrect. There is no evidence from the graph that suggests that both the tadpole and the fetal human live in a more oxygen-rich environment than do the adult frog or the mouse. Therefore choices D and E are incorrect. The curves do provide evidence that the tadpole hemoglobin has a greater affinity for oxygen than the hemoglobin of an adult frog. At 20 mm Hg, the hemoglobin of a tadpole is saturated. This does not take place in the adult frog until a pressure of about 40 mm Hg. Choice A is the correct answer.

---

155. Of the following, the last organism to succumb in an oxygen-deprived environment would probably be the

(A) tadpole
(B) adult frog
(C) fetal human
(D) adult human
(E) mouse

In an oxygen-deprived environment, the tadpole would probably be the last organism to succumb because its hemoglobin remains saturated with O_2 at lower oxygen partial pressures than does the hemoglobin of the other organisms shown. Choice A, therefore, is the correct answer.

Questions 156-160

The ability of tobacco plants to produce chlorophyll is governed by a dominant allele, *C*. The homozygous recessive condition, *cc*, produces albino tobacco plants.

In the laboratory, a student places 100 tobacco seeds on moist filter paper in each of 2 petri dishes. Both dishes are placed at room temperature for 3 days to allow seeds to germinate. Dish A is placed in the light. Dish B is placed in the dark. The student then records the data in Table 1 below.

TABLE 1

	Condition	Number of Non-Green Plants	Number of Green Plants
Dish A	Light	24	73
Dish B	Dark	96	0

Both dishes are left in the light for another day. The following data are then recorded.

TABLE 2

	Condition	Number of Non-Green Plants	Number of Green Plants
Dish A	Light	24	73
Dish B	Light	26	70

The results of the experiment indicate that the number of nongreen plants and green plants is the same whether the plants have been allowed to germinate in the light for a 3-day period or for a 4-day period. When the seeds remain in the dark for 3 days, all of the plants are nongreen. However, when the plants are moved to the light for one day, approximately 75 percent of the nongreen plants become green.

---

156. The variable that was most important in causing the results shown in Table 1 is the

(A) temperature
(B) genetic differences between the seeds in dish A and dish B
(C) amount of moisture
(D) presence of light
(E) lack of soil for germination

The seeds are allowed to germinate in the light and in the dark. It is, therefore, the amount of light, choice D, that is responsible for the result in table 1. The other factors, temperature, genetic differences between the seeds in dishes A and B, amount of moisture, and lack of soil for germination were identical for each setup.

---

157. The genotype of the non-green seedlings in dish A is most likely

(A) *cc* only
(B) *Cc* only
(C) *CC* only
(D) *Cc* or *cc*
(E) *CC* or *cc*

The homozygous condition *cc* produces nongreen (albino) plants; therefore, the genotype of the nongreen seedlings in dish A must also have been *cc*. The answer is choice A, "*cc* only."

---

158. The green seedlings in dish A would most likely have which of the following genotypes?

(A) *cc* only
(B) *Cc* only
(C) *CC* only
(D) *CC* and *cc*
(E) *CC* and *Cc*

The ability of tobacco plants to produce chlorophyll is governed by the dominant allele *C*. Thus both the genotype *CC* and *Cc* will give rise to green seedlings. The correct choice, therefore, is E, "*CC* and *Cc*."

---

159. From the results obtained in dish B, it can be correctly concluded that

(A) tobacco-seed genes spontaneously mutate under certain environmental conditions
(B) environmental factors influence gene expression
(C) gene expression is too unpredictable to be of any scientific value
(D) color in tobacco seedlings is not influenced by genes
(E) chlorophyll promotes the germination of seedlings

The seedlings in dish B all appeared nongreen in the dark; however, approximately 75 percent of the seedlings made chlorophyll when the dish was placed in the light. This was very close to the result obtained when the seeds were placed in the light for 4 days. The light, therefore, must have influenced the development of chlorophyll. Thus, choice B, "environmental factors influence gene expression," is the correct answer.

---

160. From the results obtained in Table 2, the genotypes of the parent plants of the seedlings are most likely to be which of the following?

(A) *CC* × *CC*
(B) *CC* × *Cc*
(C) *CC* × *cc*
(D) *Cc* × *Cc*
(E) *cc* × *cc*

As the ratio of green to nongreen seedlings was 3 to 1, the parent plants most likely had a genotype of *Cc* and *Cc*. The genotypes of the offspring will therefore be *CC, Cc, Cc,* and *cc*, which is a 3 to 1 ratio (green to nongreen). Hence, choice D is the correct answer.
